CREATIVE
INTELLIGENCE

CREATIVE
INTELLIGENCE

Harnessing the Power to
Create, Connect, and Inspire

BRUCE NUSSBAUM

HARPER
BUSINESS

An Imprint of HarperCollins*Publishers*
www.harpercollins.com

HarperCollins books may be purchased for educational, business, or sales promotional use. For information, please e-mail the Special Markets Department at SPsales@harpercollins.com.

FIRST EDITION

Library of Congress Cataloging-in-Publication Data has been applied for.

ISBN: 978-0-06-208842-0

13 14 15 16 17   OV/RRD   10 9 8 7 6 5 4 3 2 1

For Leslie

w.d.m.

Acknowledgments

When I was at *BusinessWeek*, I was happy to make the transition from being a voice of authority, editing the editorial page and writing cover stories, to a curator of conversations, managing online innovation, design websites, and blogs. This book puts me back into a voice of authority role, but that would not have been possible had I not also participated in many conversations about creativity, innovation, and design. It is my hope that this book will, in turn, spark many new conversations that I can participate in. Of course, I am solely responsible for *Creative Intelligence*, but I am also deeply indebted to many people for its creation.

Let me start with Ben Lee, former provost of the New School and professor of Philosophy and Anthropology. Creativity is often a product of proximity and connectivity, and had I not been given an office near Ben's when I moved to Parsons The New School for Design, this book would not exist in its current form. I developed many of the book's key concepts, especially those revolving around the social context of creativity, in wonderfully stimulating talks with Ben. His graduate work at the University of Chicago in anthropology, economics, and philosophy meshed with my graduate work at the University of Michigan in anthropology, sociology, and political science. While co-teaching courses, such as Steve Jobs and Alexander McQueen: Design as Social Movement and Design, Capitalism, and Social Movement, I honed many of the arguments made in this book. Ben has a curious, insightful, and inspiring mind, and I was just plain lucky to find myself sharing the same water cooler, literally, with him.

Parsons is a special place, drawing exceptional students from Asia, Europe, Latin America, the Middle East, and, of course, the United States. I spent a year working with a student named Kelsey Meuse before she graduated on a proposal for a Gen-Y research institute. Students from the Design and Management; Art, Media, and Technology; and Design and Communications programs have been particularly wonderful to work with. Their team projects have led to insights that are reflected in my discussions on embodiment, aura, and knowledge mining.

Special thanks go to the grad students in the MFA Fashion and Society program run by Shelley Fox. A designer herself, Shelley teaches not only the craft but also the political sociology of fashion. Carlos Teixiera, who, like me, is in the Parsons School of Design Strategies, shared with me his amazing research on the rise of new innovation and design consultancies in India, Brazil, and other emergent countries. I am grateful to Carlos for looping me into the Dream:In conference in India that shaped the way I view creativity today. And it was Carlos who introduced me to Sonia Manchanda and Jacob Mathew, cofounders of IDIOM, India's premier innovation firm, and organizers of the first Dream:In conference. By creating opportunities for those at the bottom of the pyramid to dream better futures for themselves, Sonia has become the intellectual heir to C. K. Prahalad.

Cities are the cauldron within which most innovation and creativity take place, and the dean of Parsons, Joel Towers, has helped me understand the complexities and opportunities within the urban ecosystem. Urbanism in all its forms plays a huge role at the New School, to which Parsons belongs, and there is a deep domain knowledge about cities resident on campus. It has helped sustain the work Ben Lee and I have been doing in mapping creativity in New York with our students. Jane Jacobs would be proud of the New School's approach to cities.

I first met Keith Sawyer at a conference called Creativity, Play, and Imagination at Teachers College, Columbia University, in May 2011, but one of my students had already recommended his book, *Group Genius*, months before. Sawyer is one of the giants in the field of creativity, and his research underpins many of the stories we read about the subject in the popular press. I am indebted to him for shining a light on the origins of modern creativity research and for showing the different strands of thought that flow together into the field today. In covering innovation and design at *BusinessWeek*, my knowledge of creativity came from a different source and tradition—product and industrial design. While the focus on human factors in design led to the same anthropological and sociological sources of insight as creativity, there were also differences. Keith Sawyer's work was key in helping me bridge those differences. In his work and his personal life, playing jazz piano and doing improv on stage, Keith has lived a most creative life.

I owe a huge debt to the design community as a whole for providing me with a series of mentors to teach me about innovation and creativity. My first cover story on design was called "Smart Design," and I wrote it before I knew that there was a top design firm in New York City with that name. The cofounders—Dan Formosa, Davin Stowell, and Tom Dair—have helped teach me what constitutes good design. Another cofounder, Tucker Viemeister, has taught my class about the close connections between design, creativity, and progressive education.

One of the first product design articles I wrote for *BusinessWeek* some twenty years ago was about the Woodzig, a small rotary saw designed by Ziba Design. I still use that wonderful tool, which fits perfectly in my hand. As the field of design has evolved from a simple focus on product to include a more complex concentration on strategy, Sohrab Vossoughi, founder of Ziba, has become one of the world's top advisors to corporations. Sohrab always says good design comes from having good clients, and Ziba has great clients.

David Kelley, Tim Brown, and Bill Moggridge have been key tutors in my learning about design, innovation, and creativity. David is not only a cofounder of IDEO; he founded the extraordinary d.school at Stanford. He and Tim are on the cover of what is probably my most impactful cover story, "The Power of Design." David, Tim, and Bill believed that design should go beyond the physical to include the design of experiences, services, and even social systems, such as health and education. They codified that approach into the concept of design thinking, the scaffolding upon which *Creative Intelligence* is built. When Bill Moggridge took over the leadership of the Cooper Hewitt National Design Museum, I got to know the humor and grace of the man who designed one of the first laptop computers and fathered the field of interaction design. I will miss him.

There is only one dean of a business school that I know who would say, "If you can't measure it, it might be the very most important thing about your business." That would be Roger Martin, former dean of the Rotman School of Management at the University of Toronto. Much of my work has been in the intersection of business and innovation and design, and Roger is simply the most sophisticated and provocative thinker in that space. And his critiques of financial capitalism are nothing short of heroic.

The list of extraordinary people in design and innovation that I have met and learned from includes, in alphabetical order, Paola Antonelli, Banny Banerjee, Yves Behar, Robert Blaich, Robert Brunner, Bill Buxton, Allan Chochinov, Beth Comstock,

Sam Farber, Lee Green, Ric Grefe, Ronald Jones, John Kao, Larry Keeley, Anna Kirah, Claudia Kotchka, Nick Leon, Tom Lockwood, Sam Lucente, Patricia Moore, Dev Patnaik, Chee Pearlman, Jeneanne Rae, Paul Saffo, Bob Schwartz, Craig Vogel, Harry West, Patrick Whitney, and many others.

One of the untold stories of innovation is the role *Business-Week* played in explaining the importance of design to business leaders in the United States, Asia, and Europe. Upon my suggestion and against significant opposition, editor-in-chief Stephen Shepard agreed in 1991 to support the Industrial Design Excellence Award (now called the International Design Excellence Award) given annually by the Industrial Designers Society of America and to let me write about the winning products, corporations, and design consultancies. The IDSA, under Kristina Goodrich, expanded the awards from a US-centric program to a global one. The annual design awards cover took up to fifteen pages in the magazine, and there was little direct advertising for this coverage. Over the course of a decade and a half, *Business-Week* spent millions of dollars supporting design and innovation. The business leaders that I've met over the years at the World Economic Forum in Davos considered it one of Shepard's greatest achievements in his career at the magazine.

I transitioned from print to digital media in the early aughts. My deep thanks go to Jessie Scanlon, who partnered with me in launching Innovation & Design, the first open-source online channel of news and analysis of the growing field. This team grew to include Helen Walters, Reena Jana, Matt Vella, and Jessi Hempel, a most extraordinary group of journalists. We then launched the magazine *IN: Inside Innovation*, off the digital platform of Innovation & Design, making us among the first journalists to have the skills to work in both online and print at the same time. Exciting times. This was the best team I have ever been a part of.

Thanks to the many Native American artists who have shown me how creativity and innovation really work. The artists who work in titanium, steel, silver, and ceramics have much in common with the engineers who craft jet engines and MRIs. The studios of successful jewelers and painters operate much like the best corporate and government research labs. Thanks especially to Lee Yazzie, Pat Pruitt, and Marla Allison.

Sarah Rainone did a wonderful job in organizing, energizing, and editing this project and driving it to completion. She is a delight to work with. Eli Nadeau did research and fact-checking. Lindsay Crouse dug down even further and doubled down on fact-checking. Thank you all. Of course, books need agents to get made, and Christy Fletcher is the best I have ever worked with. She understood the concept even as I was developing it. And Hollis Heimbouch, my editor at HarperCollins, knew how to take that concept to the highest level and get it out there.

I have to thank my mom for showing me that shopping is an experience, not an objective.

My wife, Leslie M. Beebe, didn't read every draft or listen to every idea that came to my mind. She played a far more significant role than that. Dr. Beebe, a professor of sociolinguistics at Teachers College, Columbia University, was the original source for key concepts in the book, such as framing and knowledge mining, which we talked about over the dinner table for years. Leslie has the most curious mind and infectious laugh of any woman I've ever known. I've had the most extraordinary journey with her since we met in the grad school library in Ann Arbor.

Contents

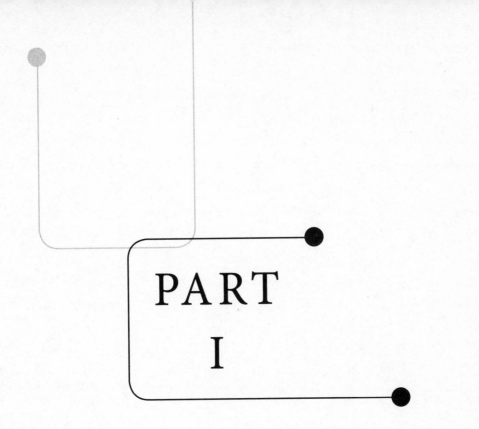

PART
I

*Reclaiming
Our
Creativity*

Strokes of Genius Are Not
What They Seem

IN 1959, A YOUNG MAN from Kent, England, was expelled
from high school after skipping assembly. "We wanted to have
a smoke," he recalled, "so we just didn't go." It was the latest in a
long line of offenses. He'd never been much of a student—even as
a boy, he'd hated school. But as a teenager, once he began to suspect that nothing he was learning in school would ever be of any
use, he "adopted a criminal mind." The way things were looking,
he was facing a future on the labor exchange.

On the surface, perhaps, the boy's behavior might have
seemed to an older generation like rebellion or apathy. And that
was certainly a big part of it—but there were other complicated
forces at work. At the time, postwar Britain was experiencing a
tumultuous period of economic and social change: British influence around the world was waning; the pound sterling was losing
its position as the global currency to the dollar; manufacturing
was in decline; the rigid class system was cracking; and young
people had begun to question their place in society.

As the country's established social order was breaking down,
a generation of young people could see that their parents' model
wasn't working. And so they were searching for new ways of living and expressing themselves, and fighting to release themselves

from the social bonds that held their parents tight in their pre-scribed roles within British society. Much the same thing was happening in the United States in the 1950s and early 60s, but the effort appeared greater in Britain, perhaps because the ties of class and tradition were ever so much stronger. "At a pace that seemed wholly un-British, various strains of unofficial culture—defiant, anti-authoritarian, and hostile to such commonplaces of tradition as modesty, reserve, civility, and politesse—were coalescing not so much in unison as in parallel" wrote Shawn Levy, in his pop culture analysis of London in the sixties, *Ready, Steady, Go!* "It was as much a revolution in English society as any the island na-tion had ever experienced," Levy said.

Yet, despite the first signs of a massive shift in the country's culture, schools still taught the old way, through rote memoriza-tion of a stale, standard, and irrelevant curriculum. "You realize later on that you're being graded and sifted by this totally arbitrary system that rarely, if ever, takes into account your whole charac-ter," the boy reflected about his education some years later. "They never took into account that hey, you might be bored."

His teenage years were not without their bright spots. Though his family didn't have a record player, the house was filled with music. Three generations shared a passionate talent for music: His grandfather recognized and nurtured his appreciation, showed him how to play his first licks and chords on a classical, Spanish gut-string guitar, and his mother knew just how to fiddle with the radio knobs. The boy devoured music by all the greats: Little Richard, Fats Domino, Elvis. And he had a talent for drawing; it was his art teacher who noticed and helped get him into Sidcup Art College.

Even at Sidcup, though, the boy didn't play by the rules. While many students studied disciplines like graphic arts to pre-pare them for jobs mostly in advertising and commercial arts, others skipped class to smoke and play music in the bathroom—it was there that he felt really alive. While classes were being held down the hall, the boy and his peers were teaching each other fingerboard work. It was around this time he became reacquainted with another boy he'd met in primary school, a young man with an identical taste in music. They spent a year hunting for records, seeking out the "real R&B" of artists like Muddy Waters and Jimmy Reed.

In 1962, the boy dropped out of Sidcup, moved into a flat with his friend, and, together, they formed a band. They started out playing rhythm and blues in London nightclubs. Though they only covered other people's songs, the boy dreamed of one day writing his own. But despite his many attempts, he had no success. Until one day when "something else took over," he said. "It was a shock, this fresh world of writing our own material, this discovery that I had a gift I had no idea existed. It was Blake-like, a revelation, an epiphany."

As he later wrote in his book *Life*, his first song was called "As Tears Go By" and was first performed by a then-unknown seventeen-year-old named Marianne Faithfull. His name was, of course, Keith Richards, and his collaborator and friend was Mick Jagger.

The partnership between Richards and Jagger would go on to form the foundation of one of the most famous rock bands of all time, one of the most commercially successful—the Stones have grossed more than $2 billion in their five decades of playing together—and one of the most lasting—even as Richards and Jagger hit seventy, they continue to play to fans around the world.

It goes without saying that Richards and Jagger are two of the most creative individuals of our time, but perhaps not for the reasons some might think.

CONVENTIONAL WISDOM ABOUT CREATIVITY WOULD tell us that the boy from Kent (not to mention his friend Mick) was a genius. That he was lucky, gifted, special—all it took was that big "aha moment" for him to discover his place in the world as an artist and creator. Americans love stories like this: stories about the mad genius, the special personality, the lightbulb of inspiration flashing on at the eleventh hour. That's one story of creativity, and there are a lot of other beliefs about creativity that go part and parcel with it.

Many of us believe that creative people are visionaries who are ahead of their time, "right-brain" types who think differently from everyone else. We've romanticized the notion of the lone poet starving in a garret or scribbling away by a pond far from civilization. We've come to have faith that science, neuroscience in particular, can explain why certain people are more creative than others, and we hope brain scan technology might offer insights about what the rest of us can do to become more creative. We may well have experienced fleeting moments of creativity in our own lives—but once it passes, we go back to life as usual, certain it was just a fluke.

Most of us experience some level of creativity anxiety. We feel that we are not creative enough, that we don't know how to be creative or simply weren't born that way. Even some of the most talented people fail to recognize that what they do is indeed creative. They aren't seeing any bright lightbulbs going off. They can't pinpoint any special moment when creativity "happened." They're not "artists"; they're just doing their jobs. I've heard of an engineer working on an advanced jet engine—basically handcrafting a gigantic, complex, high-tech machine out of titanium, successfully boosting its thrust efficiency by 20 percent—who failed to recognize that he's performing a creative activity. I've seen a student use smartphone app technology to develop a whole new way for her Gen Y friends to experience art, and still not consider herself a creative person. We don't think of ourselves as creative because we don't know how to identify creativity. We don't even know how to define it properly.

Because there is so much uncertainty about creativity—and so many myths about it—we often reject creativity in favor of predictability and conventionality, even when routine destroys our ability to enjoy our work or our lives. Researchers at Cornell University, University of Pennsylvania, and the University of North Carolina, Chapel Hill, showed that participants "demonstrated a negative bias toward creativity (relative to practicality) when participants experienced uncertainty." Worse, "the bias against creativity interfered with participants' ability to recognize a creative idea." They associated creative ideas with negative words—*agony*, *poison*, and my favorite, *vomit*. Let's face it. Creativity scares us.

As with most conventional wisdom, there is some truth to some of these perceptions about creativity—many artists do, after all, seem to move to the beat of their own drums (though perhaps that's because we don't see the other instruments accompanying them). And there have been a number of studies linking mood disorders and creative behavior, including one by Dr. Kay Redfield Jamison, a professor of psychiatry at the Johns Hopkins University School of Medicine in Baltimore, that found the rate of depressive illness among distinguished artists to be ten to thirty times more common than in the general population.

But a growing body of research is showing that other forces, social forces, are at work. Einstein's brain was extracted and examined for its shape and size, and in the end even pickled and sliced up, all ready for the microscope, because many believed the secrets to his extraordinary mathematical creativity lay in its unusual surface, density of neurons, or lack of crevices. (Lenin's brain was also examined this way, by Soviet scientists seeking to verify his genius.) Yet we know Einstein was more than just a brain. He played violin, originally at his mother's behest, but his reluctance was transformed when his discovery of Mozart inspired a lifelong love of music. He struggled in school, barely passing some subjects, while excelling in mathematics.

We also know Einstein drew groundbreaking conclusions from his work as an examiner in a Swiss patent office, where he evaluated applications for electromagnetic devices. The pace and quality of the work allowed him to dwell on recurring questions about electricity, light, matter, space, and time, which would be crucial in the experimentation leading to four of his most famous papers. While Einstein was a patent clerk, he and two friends convened a group they called the Olympia Academy, in which they read philosophers and physicists, such as Mach and Hume, and discussed their own work. Einstein acknowledged the effect of these informal convocations on the development of his thought throughout his career.

And we know that Keith Richards did not write "As Tears Go By" alone, but with Mick Jagger and only after Andrew Loog Oldham, their manager and producer, pushed them into a kitchen and told them not to come out without a song. Oldham, who'd worked for Mary Quant in London's hot fashion scene in the early 1960s and later as a publicist for Bob Dylan and the Beatles, knew how songs—and careers—got made, and told Richards and Jagger their future depended on their songwriting. It was because of their collaboration, their hard work, and this bit of wisdom from someone with experience that Richards and Jagger were able to go on to write "The Last Time," the first song played by the Rolling Stones; "(I Can't Get No) Satisfaction," the band's first hit (at least in the United States); and the hundreds of songs that came after.

That revelatory moment when the two began writing lyrics might have seemed like a stroke of genius, but it was actually a natural result of years of study and hard work: They learned to do what they love in a different kind of classroom—not by memorizing the right answers to an old test but by playing together to generate something new. "What I found about the blues and music," wrote Richards in *Life*, "tracing things back, was that nothing came from itself. As great as it is, this is not one stroke of genius. This cat was listening to somebody and it's his variation on the theme. And so you suddenly realize that everybody's connected here. This is not just that he's fantastic and the rest are crap; they're all interconnected."

Of all the words that Richards has written in his half-century-long career, these perhaps are the most apt to begin a discussion about creativity. It's a discussion that's a little different from the ones we're used to having, and it's one whose time has come.

Richards's story reveals that we all have the capacity to be creative. We just need to search a little deeper to bring it out. Indeed we have been searching for the source of creativity for some time.

The Search for the Secrets of Creativity

I BEGAN COVERING DESIGN AND innovation in the 1990s for *BusinessWeek*, and during my years at the magazine, I was involved with many attempts at measuring innovation. In 1992 *Business-Week* began supporting what is now called the International Design Excellence Awards (IDEA), highlighting new innovations at corporations and consultancies in the United States, Europe, and Asia. In April 2006, the magazine teamed up with the Boston Consulting Group to launch a list of "Most Innovative Companies" using an algorithm that combined soft and hard data—the collective opinion of senior managers as well as metrics on R&D spending, patent listings, and revenue generated by new products.

Then, in 2008, my team created an S&P/*BusinessWeek* Global Innovation Index, which consisted of twenty-five global companies that were considered "creativity-driven" and whose stock prices over time were higher than the average. The goal of the index was to facilitate investing in innovation within big global corporations, and we used conventional measures for ranking—R&D spending, number of patents, revenues generated from new products.

Between our many lists, the index, and in 2006 the launches of the online Innovation & Design channel and *IN: Inside Inno-*

vation, a quarterly magazine filled with insights and advice on innovation, we put in a massive amount of work and effort to covering innovation over the years. We wrote hundreds of stories about innovation successes, compiled scores of lists, and congratulated ourselves on really covering innovation.

And then I stepped back and really looked at the results.

In list after list of "Most Innovative Companies," and year after year at the IDEA annual awards, the same two dozen or so companies showed up time and again. Thousands of public and private companies launch innovative products and services each year, but only a tiny sliver of that group ever showed up.

To add insult to injury, the one company that everyone agreed was the *most* innovative didn't fit any of traditional measures; that was, of course, Apple. Apple spent very little on R&D; it didn't have a formal innovation "funnel" process with established procedures; it made just a few things and it was run by an imperious CEO and a small band of followers. Here was this company that was transforming our lives—not only by giving us beautiful, elegantly designed products, but by changing the very way we interacted with products, giving us the tools that allowed us to create everything from playlists to photo albums to movies.

Apple simply didn't fit. And that wasn't the only important omission. When you look at most of the products that have changed our lives over the past decades—from Facebook to Twitter, Amazon to eBay—they are almost invariably start-ups offering surprising new products or services. Their success had nothing to do with the number of patents or the amount spent on R&D, and so it was impossible to measure these companies using the same metrics of more established organizations. That made it clear that something else was happening, something else was responsible for these big disruptive innovations, something that we hadn't yet discovered how to quantify.

Frustrated, I began asking around, starting with my contacts at a number of consultancies whose job it was to help existing companies become more creative. They were reluctant to talk about it at first, but eventually they began to share the same surprising truth: They had an incredibly low success rate when it came to helping companies usher in the kind of transformative innovation they were seeking.

We've all read about the many corporate innovation successes (*BusinessWeek* published articles about many of them, from Kaiser Permanente's shift from focusing on medical treatment to the "patient experience" to P&G's development of the Swiffer brand), but few of us are aware of the large number of failures. I prodded people and, grudgingly, they agreed that innovation was hard to come by. Years after I initially began these conversations, the heads of two top innovation firms would tell me that of the hundreds of projects they worked on each year, only a handful actually worked.

After spending most of my career covering innovation, developing tools to help companies and investors measure it, and believing that companies really could become more innovative, I was shocked. And every bit as frustrated. What was this all about? Why were all the hugely disruptive innovations coming out of left field? Why were companies that were spending money and time on all the right things failing to come up with the same kinds of life-altering products and services that some twenty-somethings could do with zero budget? How could the millions spent on metrics and measuring by big organizations generate far less innovation than some talented computer programmers and their buddies?

Sure, you still saw significant creativity coming out of the government—the Internet and the technology that led to Apple's Siri are two excellent examples—but the closer you got to what was going on there, the more it seemed that the government labs had less in common with Fortune 500s than they did with start-ups. In fact, I began to dig into what actually happened inside the best government and corporate labs, only to discover the social dynamics that led to innovation—serendipity, connection, discovery, networking, play—mirrored the organic messiness of a great creative city or college campus more than the mechanical process of a big corporation. So what was going on?

I began to ask myself not only "What are we measuring?" but also "What *aren't* we measuring?"

CRACKING THE CODE OF CREATIVITY

I wasn't the only one asking these kinds of questions. By the middle of the first decade of the twenty-first century, design thinking had become wildly popular. Ironically, it was during a discussion about design thinking and innovation that some of us came to see the limits of focusing on design alone.

I had joined Danny Hillis at his Applied Minds studio in Glendale, California, on October 30, 2007, along with the big guns in design, design thinking, and innovation—Larry Keeley from Doblin, Roger Martin, dean of the Rotman School of Management at the University of Toronto, and Patrick Whitney at the IIT Institute of Design in Chicago. Hillis, a world-class innovator who invented parallel computers and worked for both Disney and the US military (secret stuff), wanted to talk more about the process of innovation. He knew that he was good at generating new ideas, but he didn't know why, and so he had no idea how to pass this ability to his staff. "How do you make innovation routine?" he asked. Each of us gave our thoughts on the subject, but it was Craig Wynett, then a senior innovation manager at Procter

& Gamble, whose answer changed my entire outlook on the subject. Yes, he said, P&G under CEO A. G. Lafley had increased its innovation by pulling down the silos separating scientists, engineers, and designers. "But," he added, "that's only 5 percent of what needs to be done. We really need to deconstruct the creative act." Wynett was saying, in effect, that we needed to crack the code of creativity.

His insight was a game-changer. Wynett was saying it wasn't enough to just remove the boundaries and "unleash" innovation. We had to better understand how creativity really worked. It transformed my whole perspective about how to "do innovation right." While most Fortune 500s were no longer driving innovation, P&G was. With products like the Swiffer mop and the inexpensive SpinBrush electric toothbrush on their roster, P&G under A. G. Lafley had excelled in innovation for some time. Wynett was Chief Innovation Officer, but he wasn't talking about innovation, and he wasn't talking about design.

He was talking about creativity.

Over at the Stanford "d.school," known formally as the Hasso Plattner Institute of Design, IDEO founder David Kelley had begun a similar conversation. Though a longtime champion of the process of design thinking, by this time, he, too, was talking about creativity. "Creativity is like a foreign language," Kelley told me. "You need two abilities to be competent in the world today. You need analytic ability and the tools that go with it. And you need creative ability and the tools that go with it." The d.school was even giving Exec Ed–type courses to managers to build up what Kelley was describing as "creative confidence."

As focused as I was on design and innovation at the time, I couldn't ignore the wisdom in what Wynett and Kelley were saying. I began to think of creativity as a wider, richer, more accessible concept than design. I started discussing this with business leaders and writing about creativity for *BusinessWeek* and found an extremely receptive audience.

In 2010, IBM ran a survey of 1,500 CEOs and found that the most valuable management skill was no longer "operations" or "marketing" but "creativity." By that time, I had joined the School of Design Strategies at Parsons partly in search of better answers to my questions about creativity and partly in search of better questions. It was in my new role as Professor of Innovation and Design in the School of Design Strategies that I was able to step back and see that I had not connected enough dots when it came to truly understanding and assessing creativity or innovation. We weren't explaining this at all, and in fact our whole focus was misplaced. What we needed to do was widen our lens—not only by looking beyond the traditional measures of success, but also by looking back to the history of capitalism, to the first thinkers writing about the relationship between creativity and capitalism.

Meanwhile, I began to look at the challenges facing my students at Parsons. They had parents advising them to get jobs at big corporations because they offered stability. But my students knew better: They could see that the world was in a state of flux. They knew young people just two or three years older who had already had three or four careers—not just different jobs but jobs in entirely different industries—by the time they were twenty-five. My students could see that they needed to create their own opportunities, quite possibly their own companies, if they wanted to succeed. They understood that they needed to transform themselves constantly in order to survive, and they wanted to learn how.

Listening to them talk about the challenges they faced pushed me to think about creativity as something you might train for, as a skill that could be assessed.

I was fortunate to find out that I wasn't alone. In March

2010, I joined a remarkable group of design thinkers at Stanford for the Future of Design Conference, organized by Banny Banerjee, then-director at the Stanford Design Program.

The goal was to go beyond the design thinking paradigm that worked well enough but many at the conference felt just didn't scale. Given the tumultuous times, scale was critical if design was going to have an impact. In the intense discussions among us that took place in those two days, some new terms emerged—"Design Intelligence," and "CQ" or "Creative Intelligence." Who coined these phrases first remains shrouded in the collective memories of the participants. In the end, we all agreed that however you named it, assessing this kind of intelligence was very important. But how do you do it?

As it turns out, a number of people had tried.

ON A COLD WINTER NIGHT at a safe house in Manhattan somewhere near Bloomingdale's, CIA Director William Casey told me about the CIA's methods for recruiting spies in World War II. It was 1983, and I had received a call the week before from a woman claiming to be the assistant to William Casey, the chief of the CIA. Thinking it was a joke by a friend, I laughed and laughed, until Casey cut in with "Hiya, Bruce, I just finished your book. Can we talk?" That book, *The World After Oil: The Shifting Axis of Power and Wealth*, analyzed how nations would respond to the rise of computer technology. I predicted the Soviet Union would have serious difficulty and the Soviet empire could fall apart. This was the height of the Cold War, and it piqued Casey's interest. He got on the phone and I hurriedly arranged to meet him.

Casey had been head of the CIA for two years; prior to that, he'd successfully managed Ronald Reagan's presidential campaign. But Casey's career in espionage had begun forty years earlier when he, then a Navy lieutenant junior grade, was summoned by William "Wild Bill" Donovan, the legendary "Father of American Intelligence," to join him at the Office of Strategic Services (OSS).

Donovan immediately liked Casey. He could recognize aspects of himself in the young man; both were descendants of poor Irish immigrants, successful lawyers, devout Catholics, and fervid Republicans, and Donovan recognized the same kind of "restless, devouring mind that leaped from enthusiasm to enthusiasm. . . . He was immune to conventional patterns of thinking, preferring to rely on his intuition."

Despite his talents, Casey still felt like an outsider in his first days at the OSS, an institution dominated by what he called the "white-shoe boys," upper-class Wall Street types, rich scions of famous industrialists like Alfred duPont and socialites like Junius and Henry Morgan. Determined to prove himself, Casey left for London in November of 1943, leaving behind his wife and daughter.

In December of 1944, the thirty-one-year-old Casey, now the chief of SI (Secret Intelligence) for Europe, was charged with recruiting spies to gather information inside Nazi Germany. Strangely, this had not been done before. While agents had been sent into occupied Europe—France, Italy, Belgium—where local partisans on the ground could help them, there was no network of partisans inside Germany itself. It was considered too dangerous.

After Germany's surprise offensive at the Battle of the Bulge, however, the Allies changed tactics. With the end no longer in sight, they needed information inside Germany to help them, finally, stop Hitler. Casey's SI agents were given several tasks: They were expected to gather information on potential targets for the Eighth Air Force to bomb, especially German troops massing at rail centers; compile statistics on German industrial production, analyzing whether heavy bombing was slowing it down; monitor progress on the development of "wonder weapons," including rockets and the atom bomb; and finally, confirm the existence of an "Alpine Redoubt," the Nazi's much-whispered-about bunker complex rumored to be located somewhere in the mountains of Bavaria.

Casey picked his agents from a pool of German communists and labor organizers who'd escaped to London as well as Poles and other Eastern Europeans who could pass as foreign workers, according to Joseph Persico in his book *Casey*. Four decades later, Casey told me, in his thick Long Island Irish accent, exactly what they were seeking in their prospective agents. They wanted people who were smart, not afraid of risk, and independent, and they hired many psychologists and psychiatrists to test for those traits.

What the OSS was looking for was, in essence, creativity. According to one report from the Joint Special Operations University and the OSS Society Symposium, "what made an effective OSS direct action operator was a secure, capable, intelligent, and creative person who could deal effectively with uncertainty and considerable stress." OSS trainees were encouraged to "use their own ingenuity and creativity in overcoming problems."

The night we spoke, Casey was sitting back, reclining deep into a sofa. It was late in the evening, he'd taken off the jacket of his three-piece suit, and his vest was covered in peanut shells. He held the peanuts in one hand and a Scotch in the other. He'd had several. I asked him what happened to the people who were parachuted into Europe.

They just kept disappearing, he said. Staring ahead, perhaps thinking about specific people and events that had occurred some forty years earlier, Casey didn't elaborate further. We now know that many spies did succeed. Moe Berg, a Columbia Law School graduate who also played major league baseball, reported that Nazi Germany was not close to building an atom bomb. The actor Sterling Hayden made key connections for the OSS running supplies and information through Yugoslavia and Fascist Italy. But my talk with Casey did make me wonder why the psychological testing for ingenuity and creativity didn't stop some good people from disappearing behind enemy lines.

It would be thirty years before research would prove that scores on personality tests didn't correlate with real-world creativity in the field.

THAT NIGHT, CASEY WASN'T SIMPLY telling me about OSS recruitment methods in World War II; he was also telling me the story of the origins of America's search for the secrets of creativity—a search that continues today.

In his book *Explaining Creativity: The Science of Human Innovation*, R. Keith Sawyer, a professor of psychology and education at Washington University in St. Louis, traces the history of modern creativity research back to World War II, describing how former OSS and military psychologists went on to launch research institutes that studied creativity at UC Berkeley, the University of Southern California, and the University of Chicago. One of these men was J. P. Guilford, who would later develop one of the most popular and extensive creativity tests, which initially measured 120 personality traits, including originality and flexibility.

During the Cold War, federal funding for creativity poured forth. When Harry Truman set up the National Science Foundation in 1950, according to Sawyer, one of the first projects was to identify the most promising future scientists. The NSF funded a series of key conferences at the University of Utah on the identification of creative scientific talent.

By the sixties, the search for creativity had spread beyond the university to the nation's public school system. Testing children to identify those with the potential for creativity in order to steer them toward careers in science and technology became a key goal of creativity research. In 1960, Ellis Paul Torrance developed an exam that essentially tested for "divergent thinking"—the ability to come up with many potential answers to questions, not just the right ones (which traditional IQ testing did). Torrance's exam is still the most widely used to test for giftedness and creativity in both children and adults. It has been translated into thirty-two languages and is the basis for more than two thousand studies.

But despite the widespread use of these creativity tests, researchers in the 1970s and 80s began to challenge the assumption that testing for mental abilities and personality traits could really predict future real-life creative behavior. Some raised questions about sampling. Others expressed doubt as to whether high scores on these tests translated into real-world creative output.

In the eighties, Teresa M. Amabile, a Stanford PhD in psychology with a focus on creativity, went back and looked at nearly all of the personality tests that measured an individual's "originality" and observed that there was an implicit subjective bias built into the tests. As Sawyer points out, Amabile concluded that people from different fields and careers have their own measures of creativity and novelty and were therefore scoring tests differently according to their own areas of expertise.

Amabile's work is significant for two reasons. She argued that creativity has a social context; each field—whether music, business, science, sports, art, or warfare—has a different set of experts who have specific notions of what is traditional, conventional, and creative. So while there may be general patterns of creative behavior that everyone shares, creativity in the field requires a certain level of domain knowledge.

Amabile's research also marked the beginnings of business's love affair with creativity. After her breakthrough research, Amabile shifted her focus away from individual creativity to organizational motivation and team creativity. She is now a director of research for Harvard Business School, ensuring that the business school supports creativity research as it emerges from within a social and cultural context, such as a new start-up, a project team, or even an established global corporation. This particular thread of creativity research—that moves away from the individual to the group, from personality and thinking patterns to social organization and behavior—has only gotten stronger as business leaders express intense interest in how to make their organizations more innovative.

By the time Amabile and others began to critique the first wave of creativity research, Sawyer notes, a second stage—one that has its origins in the personal growth movement of the 1960s and 70s—was growing in popularity. As cognitive psychology was gaining traction, psychologists sought to explain creativity by showing how it emerges from mental processes and abilities. Where before psychologists tried to find out how personalities and individual thinking differed, now they looked at what mental processes people shared and how they correlated to—if not caused—specific behaviors.

It was in the labs of Lake Forest College and later, in the seventies, at the University of Chicago that one of the most important contributions to modern creativity research was developed: "flow." The experiments of psychologist Mihaly Csikszentmihalyi determined that there is a distinct cognitive state of mind that individuals enter when they are performing creatively. We talk about being *in the zone, on a roll, centered,* or *in the groove,* but my favorite description of this state of mind just might be Keith Richards's description of playing with the Stones: "There's a certain moment when you realize that you've actually just left the planet for a bit and that nobody can touch you. . . . When it works, baby, you've got wings."

At Chicago and in a series of books including *Creativity: Flow and the Psychology of Discovery and Invention*, Csikszentmihalyi laid out the elements that went into the flow state that Richards describes. The absence of a sense of time, extraordinary focus, a feeling of unlimited potential, great confidence, intrinsic motivation, absence of hunger and fatigue, and joy, even rapture, in the process of creating are all elements of flow. Back in 1959, Rollo May, an existential psychologist, used the term "peak experience" to describe moments of intense awareness, heightened consciousness, and an obliviousness to time and surroundings. Csikszentmihalyi built on these ideas and called it the "flow state."

Csikszentmihalyi gave a secular, psychological interpretation to what artists for centuries have described in religious terms such as "God's gift." Even today, many traditional potters, weavers, and jewelers on the Hopi, Navajo, and Pueblo reservations talk about the moment when, after days or weeks of feeling blocked in their work, a "spirit" from the Creator moves through them, providing insight and the path to new creativity. They often describe this moment with arms raised toward the heavens but use precisely the same terminology of timelessness, insight, potential, and joy.

By the 1990s, advances in brain imaging technology were allowing psychologists to directly see what was happening in the mind when people were performing various "creative" tasks. As powerful scanning machines that generate three-dimensional images of the brain at work were developed, the new field of cognitive neuroscience was born. It continues to be a well-funded and popular research methodology for analyzing creativity today.

Cognitive neuroscience research has helped demolish a number of major myths about creativity. Brain scans have shown that creativity is not localized to the right side of the brain, despite the popular perception about the creativity of "right-brain" types. Because creative behaviors activate the entire brain over a period of time, creativity can't be reduced to a single flash of insight in a single moment. And when scanners get smaller and more complex, we may discover more insights into what happens to the brain when we act and feel.

But even then, neuroscience runs the risk of being just like the shadow on the screen in a Javanese puppet play, revealing the reflection of reality, rather than the reality itself.

And so, as fascinating as the new research on creativity from neuroscience is, as much as it has helped to debunk the idea of the lone genius, it's time to also toss out the old lightbulb, and turn a more wary eye on the brain wave machines that so beguile us. As cool as "aha moments" are, and as interesting as it is to understand what parts of our brain are working when we're improvising or solving a problem when we're in the shower, creativity is about so much more than that moment . . . and it's about so much more than the individual experiencing that moment.

Until neuroscience can stimulate parts of the brain and cause us to behave creatively—and it's well worth asking ourselves, at what cost?—then I'd argue that the best way to understand Creative Intelligence is to study and learn from the people and organizations who've cultivated it.

Not surprisingly, several of the pioneers of modern creativity research have done just that.

IN RENAISSANCE ITALY, FLORENCE HAD a remarkably large number of great painters: Leonardo da Vinci, Raphael, Botticelli, and Donatello among them. How do we explain it? Was it serendipity that there were so many people with peak "flow" states that enabled them to create? Perhaps. But there may be another explanation.

Florence at that time was an immensely rich city-state whose elite marked their status by acquiring art. The presence of great artists tended to generate more great artists as they frequently collaborated, seeking each other's criticism and advice. In other words, the creativity emerged in large part from the social and cultural context of Florence six hundred year ago. By the measure of the Renaissance, great painting and creativity do not thrive in Florence today and may never again. The right mix of political and economic conditions that made them possible no longer exist.

Interestingly, the man who observed how the culture of Renaissance Italy contributed to the explosion of artistic creativity was the same man who gave us the psychological concept of individual "flow"—Mihaly Csikszentmihalyi. Though Csikszentmihalyi is much better known for his work on flow, in four works of research, he returned again and again to the lives of artists during the Italian Renaissance of the fifteenth century. In his research and his books, Csikszentmihalyi asks: What are the social conditions that lead to creativity? How can we make ourselves part of that social matrix? These are typically questions that anthropologists and sociologists, not cognitive psychologists, ask. They involve the larger concepts of culture, organization, change, and social movement.

It was Csikszentmihalyi's student R. Keith Sawyer who highlighted that these two streams of creativity research—one focused on cognition, the other on culture—were pioneered by one man. Csikszentmihalyi has had many students over the years, but few have had the same impact on creativity as Sawyer.

Sawyer grew up in Newport News, Virginia, and received a computer science degree from MIT in 1982. As a student, he used what was then a new field, AI, artificial intelligence, to simulate in computers what was assumed by his professors to be the creativity of humans. But all the experiments were designed on the assumption that creativity was an individual process. That didn't fit what Sawyer already knew from his days playing in a jazz band. "I saw creativity as it emerged within an ensemble between people, not just what was inside someone's head," he says. "AI missed the interactional dimension where creativity takes place." This "interactional dimension" is something many artists have experienced firsthand. In *Life*, Keith Richards writes, "It was the bands behind them that impressed me just as much as the front men. . . . It was how guys interacted with one another, natural exuberance and seemingly effortless delivery."

With his degree, Sawyer became a consultant, designing video games for Atari and, during most of the nineties, working as a management consultant on innovative technologies for Citicorp, AT&T, US West, and other global corporations. While pursuing a degree in psychology and education at the University of Chicago, he took a course called the Psychology of Creativity with Csikszent-mihalyi in which he observed the conversational dynamics of young children doing what he called "sociodramatic pretend play." This interaction reminded him of his own music. "It was just like what jazz players do," he says. "I played jazz piano since being a teenager and continued to play in college and after college. The thing about jazz is that it is an ensemble art form. So much of what happens is between the musicians, not inside any one musician's head."

Sawyer is now a professor of psychology, education, and business at Washington University in St. Louis, and may soon become dean of the business school's executive MBA program in Shanghai. He advises companies on how to boost their creative output, and his advice comes directly from his experience with children pretend-playing. "What makes for a great creative team? Whether it's musicians, improv acts, or business teams, there are three elements to creative teams: trust, familiarity of members with each other, and a shared commitment to the same goals," he says. "These can enhance the performance of any group."

If cognitive psychology and neuroscience have taught us that we all possess the ability to be creative, then a more sociocultural approach offers insights as to how we must *act* in a social context to *be* creative. How does creativity emerge from collaboration, how does it thrive within a social context? In an era of huge social change and the explosion of social media, it's *the* question to ask.

It is not known how many of the nearly 24,000 OSS agents who worked for the OSS during World War II succeeded—or even survived. Many did exceptional work. But as I talked with Bill Casey in those hours on that night, it was clear that many did not. What I believe Casey was telling me was that he had learned that success as a spy had little to do with a trait or personality type but rather a particular kind of training: The spies who lived to tell the tale of their service in Europe were people who'd been trained to be creative on their feet.

A number of them—not surprisingly, perhaps, with what we now know about creativity—were trained for the stage, trained to play. In the end, he told me, the people who did best, who tended to disappear less, were often actors and sometimes clowns. The OSS sent the players in, and they tended to disappear less, he told me. And then he laughed.

Working in an environment of extreme uncertainty, improvising solutions to unexpected and often unknown problems, seeing connections between seemingly disparate data and events— these are just a few of the challenges that unite the OSS spy experience with what many of us face today. When society is safe and stable, what we need most are the tools to make things a bit better, more efficient. When technological, political, and environmental shifts threaten the status quo, what we need most are the tools to make things sharply different, radically new. We need to be less incremental and much more creative.

And yet, the prevailing view on creativity is psychological, mental, brain-centered, individualistic. We tend to believe that creativity comes through the individual and is expressed solely by the individual. These perspectives on creativity don't explain how to act creatively, how to learn to be creative, or how to assess the performance of creativity; they explain the mental processes associated with creativity.

We need to go further. We need to stop searching for some magical place in the brain where creativity resides. We need to believe in our own abilities to create and to improve upon our creative skills by teaming up with the right people. We need to stop studying creativity just in labs—and recognize that it's all around us: in the stories of great painters and their rivals, in the meals we cook using a bit of one recipe and a bit of another, in the games we play with our kids.

As for those of us studying innovation and working in organizations who are seeking to find "the next big thing," we need to explore creativity not at the individual level but as it plays out in groups and cultures. We must be willing to have deeper conversations that go beyond the standard metrics.

For some time, those of us in design thinking circles have attempted to "talk the language of business." And as we do so, the process of creation has often been reduced to a linear, mechanical process of funnels and inventories and inputs and outputs, all measured meticulously and continuously. The result, often, was incremental change—certainly good, but definitely not disruptive.

Disruption comes most often from entrepreneurs and startups whose founders were themselves swept up in cultural change and social movement. What we're witnessing in tech hubs like Silicon Valley or amid the growing start-up scene in New York isn't all that different from what Csikszentmihalyi described in his writings about Renaissance Florence. Although many of today's innovators are working on different canvases, the factors contributing to the explosion of creative output—a spirit of collaboration and competition, wealthy investors keen to be part of the next big thing—are similar. And like the artists who came before, today's innovators take what money can't buy—the desire to share among those who are lonely, the drive to participate among those who refuse to be passive, the need to build from those who don't simply want to consume—and transform it into products and experiences that people *can* buy.

And so those of us in the design and innovation world need a new vocabulary and a new repertoire of methodologies. Ideas like "user focus," "visualizing," and "failing fast" were all healthy attempts to understand and implement innovation, but we now know their limitations. "User experience," for example, was a brilliant turn away from the prevailing technology-centric, engineer-driven approach to designing products that drove so many of us crazy—and a turn toward a focus on what users needed. But why the focus on just *needs*? Why not, as Indian design firm IDIOM's cofounder Sonia Manchanda has proposed, aim for something bigger, richer? Why not focus on aspirations—dreams that we may believe are not even possible?

For decades, brainstorming, a technique developed by the advertising agency BBDO that took off in the fifties, has been perhaps the single most popular technique for generating creativity. People gather in a room and offer up dozens, if not hundreds, of original ideas. Brainstormers are encouraged to "go wild" and "think outside the box." But as Sawyer points out in his book *Group Genius*, there is a large amount of research that shows the technique doesn't work. Throwing out hundreds of ideas doesn't necessarily lead to the right one. People often hold back their best ideas in groups; many generate more ideas when they're alone. And perhaps most important, the most meaningful ideas typically result from deep domain knowledge and expertise, not from the shallow, breathless explosion of scattered thoughts.

It's clear we've reached a point where the older categories are coming under attack for not delivering all of what they have promised. And so it's time for new categories and a new framework for understanding and implementing creative innovation.

Where the old model focused on managing innovation as a process, the new model concentrates on how entrepreneurs, artists, scientists, designers, engineers, and the rest of us can transform our creative ideas into creations that have value. Where the old model focused on meeting needs, the new model gets at the heart of what is truly meaningful to people. Where the older model sought to make innovation predictable and risk free (the *Harvard Business Review* ran a cover story in May 2012 entitled "Innovation for the Risk-Averse"), the new model sees creativity as a practice that actually harnesses uncertainty. Uncertainty, after all, is where you find opportunity.

What I have discovered from twenty-five years of writing about innovation and creativity, from the hundreds of interviews I've conducted with leaders in the fields of business, design, and technology, is that there is nothing "rare" about creativity; it is something we can all cultivate. Creative Intelligence can be found across many fields and disciplines, in all spheres of life—people who might never consider themselves "creative" are drawing on many of the same skills as those a musician or writer would use. Most important, Creative Intelligence is social: We increase our creative ability by learning from others, collaborating, sharing.

You can't "do creativity" in a vacuum—and even if you could, these days, you simply can't afford to. We're living in a time of instability and immense change, and creative collaboration is key. There are many forces of change washing over our lives today, all of which require the opening up of silos, the mixing together of the incredibly specialized knowledge many of us have come to possess, the sharing of ideas across cultures and generations.

Those of us who grew up during an era of American ascendancy can recall when the culture of the United States in all its forms—music, business, style, money, language—dominated the world. But today, by most economic measures, the United States is falling relative to Asia and much of the rest of the world: China's influence continues to grow, Indian consumers demand their own style of consumer products, and Washington has to curry favor with foreign governments to buy soaring US national debt. The United States must accommodate its products and its policies to this increasingly complex world.

At the same time power is shifting throughout the world, the United States is experiencing a significant shift within its own borders. The aging of the huge boomer generation, the largest of its kind in US history, and the rise of an even larger demographic, Gen Y, marks a dramatic split. Gen Y's stronger embrace of sustainability, same-sex marriages, and racial and ethnic integration as a result of having been born into a world where such cultural values are more widely accepted is affecting not only those born after 1980 but the nation as a whole. But it's Gen Y's desire to participate in and create their own media that's having the most profound effect on the American economy.

Social media's rise has drastically altered industries from journalism to health care, altering virtually everything we do and how we do it. Facebook, Tumblr, Groupon, and Spotify all use technology that allows people to directly build their own communities and organize in flat, horizontal, and more democratic ways. This is an extreme departure from the traditional hierarchical, centralized, authoritarian modes of organization of decades past. Corporations, schools, hospitals, and virtually all large organizations need to adapt to social technology—or be replaced.

In 2010, for the first time in human history, more than half the people living on earth lived in cities. In forty more years, it's projected there will be around 9 billion people on earth, the majority of whom will live in cities, striving for a 1990s US lifestyle, eating lots of meat, living in big houses or apartments, and riding around in cars. It is a wonderful picture of upward mobility for hundreds of millions of people and a valid goal for policy makers, but it will weigh heavily on the earth's resources.

The US military's term for the kind of environment is VUCA—"Volatile, Uncertain, Complex, Ambiguous." The acronym came out of the US Army War College in the late nineties to describe the new operating conditions that world military leaders had to face—the rise of terrorism, global political instability, and asymmetrical warfare. It's a term that also perfectly captures the prevailing instability of society in general.

While the risks the average American faces certainly are less deadly, I've found no better way of describing the current economic landscape than VUCA. Change in life is a constant, but shifts in our industries or career paths usually come episodically, giving us time to adapt to them. Sometimes, however, the frequency and volatility of change happen at an unusual rate—that's where we find ourselves today. While it can be scary, we should take note from the military that VUCA landscapes also present unusual opportunities to do things differently. It will take not only new strategies, but a new way of thinking, communicating, and creating.

We were trained to deal with a world of predictable futures, but the future—both the good and the bad—is anything but predictable. We're living in an "I don't know" world where we can't fathom the problems to come, much less the answers. Skills once perceived as valuable, degrees considered prestigious no longer guarantee job security or even a middle-class income. Many of today's most in-demand jobs didn't even exist a decade ago. We need to prepare ourselves for jobs that don't yet exist, using technologies that haven't been invented, to solve problems that we haven't recognized.

And so this is a book for people who aren't just interested in becoming more creative, but who want to create things that change our lives. The five competencies of Creative Intelligence aren't simply best practices for organizations to transform themselves; they are tools that can help you plot a career path if you're young and transform your career if you're not. Understanding them will help you better navigate a rapidly changing world and construct a place for yourself in it. The competencies can also help us generate the kind of jobs, businesses, and revenues that the nation so desperately needs today.

THE FIVE COMPETENCIES OF CREATIVE INTELLIGENCE

KNOWLEDGE MINING. The knowledge at the foundation of Creative Intelligence is not the kind that can be found on standardized tests. Today's most creative entrepreneurs, thinkers, and artists are in touch with what's truly meaningful to people—starting with themselves. They understand that what matters to one generation or demographic may mean little to another. They don't focus on "unmet needs" when it comes to developing new ideas; they use their own experiences and aspirations as a starting point for dreaming up new companies and technologies. When their own experience is insufficient, these individuals don't turn to traditional market research; they go straight to the source and partner with people who are more embedded in a culture than they are.

The people who are routinely creative are skilled at connecting information from various sources in new and surprising ways. They know how to cast for new ideas—bringing together information from different fields or going back in time to discover forgotten ideas and practices they can use to meet new challenges. And there are those of us who have such deep domains of knowledge that we can intuitively understand what's not there.

In this chapter I'll introduce strategies that some of the world's most creative people use to draw inspiration from their own experiences as well as from the unlikeliest of places.

FRAMING is a focal lens that can guide us through the vagaries of a volatile world. Understanding your frame of reference—your way of seeing the world as it compares with other people's—is a key strategy no matter your aspirations or industry. People who understand framing techniques are better able to shift their perspectives depending on the situation, environment, and community they're interacting with. This is not to say that they lose sight of their aspirations or what's meaningful to them—quite the contrary. They are able to continually "check in" to see how their biases might be affecting the conversation or how their worldview might be limiting their ability to come up with more creative strategies. The concept of framing has its origins in sociology and anthropology, but I have come to see how creative individuals excel when it comes to framing their world and interactions.

As we witness the crumbling of even the most established financial and corporate institutions, framing has become an essential tool for adapting quickly to unexpected shifts. For example, in the past, when people thought about health care, they focused on treating diseases. Today the Mayo Clinic and other leading health-care facilities are focusing on well-being. In the field of education, Stanford and other top universities are beginning to utilize techniques such as search and networked delivery of information so that people can learn anywhere, anytime.

I will introduce three kinds of framing: Narrative Framing, which is how we interpret the world; Engagement Framing, how we interact with each other; and What-If Framing, how we imagine the unthinkable to innovate beyond our wildest dreams. Understanding how to frame (and reframe) our beliefs about our organizations and entire industries is a powerful way to drive disruptive creativity. At every step of the creative process, people who understand the power of framing are able to recognize where they stand, when they need to refocus their lens, and who else needs to be in the picture.

PLAYING is not just kid stuff; it's a complex behavior that is driving the creation of life-altering technologies and companies. Creativity can be found in many kinds of "playgrounds"—spaces (not necessarily physical) where people are given permission to play games, make up new rules, discover different ways of winning. We associate playgrounds with children, but Navy Seals, scientists, and engineers all "play" at discovering solutions to challenges, some of them deadly.

It has become fashionable in innovation circles to talk about the "need for failure" to achieve something great. But why label such an essential step in the creative process "failure" at all? By adopting a more playful mind-set we're more willing to take risks, explore possibilities, and learn to navigate uncertainty, without the paralyzing stigma of failure. Moreover, new research is showing that playing can be a superior alternative to a problem-solving approach to innovation.

Games are the fastest-growing social structure in society today. A generation raised on multiplayer video games is using their experience to create new business models in the fields of finance, education, sports, manufacturing, medicine, music, and art. Many companies, upon discovering that fun and competition are excellent motivators, are using games in their hiring process. And perhaps the most exciting development is the number of organizations experimenting with games for the social good.

Playing games and learning how to design games effectively teaches people not only how to create new products and services, but also how to build their own complex social systems. Gamers build communities rather than simply rack up customers. Games are dynamic, interactive, and immersive and can have any number of solutions or conclusions. For generations brought up in "Search" mode, games are the perfect organizational structure for learning.

MAKING is the fourth Creative Intelligence competency, and it's perhaps the most surprising and exciting shift to arise in our global economy. After decades of rewarding mental agility—trading on Wall Street, consulting, strategy, and branding in Corporate America—we are experiencing a maker's renaissance. Americans want to make things again. And thanks to a whole host of new technologies and the democratization of the tools of creativity—from Photoshop to 3-D printers to Behance—we're doing it.

The revival of a "maker culture," combining open-source philosophy, new channels for distribution made possible by social media, and a shift to DIY, Made-in-the-Hood consumerism, has helped Making become a critical component of innovation once again. New forms of community-curated venture capital, such as Kickstarter and Grind, are making it easier than ever to get financing for a new endeavor. This chapter reveals how we can all learn some crucial twenty-first-century maker skills, and, in so doing, re-create our jobs, careers, and identities.

PIVOTING from the inception to the production side of creation is the final of the five competencies. Traditional notions of creativity separate the process of coming up with new ideas from the actual making of new things. But truly creative people don't stop at the idea; they make the pivot into creation. By moving beyond the creative idea in order to create new products and businesses, Pivoting is a way of reprising creativity's crucial role in capitalism as a driver of innovation and growth. But how?

Most disruptive innovations come from individuals who are leading a cause and who've inspired a loyal following to get involved in that community. And yet our investments into innovation don't always reflect that. Most of our efforts to promote creativity go to older, established corporations, where incremental innovation is, at best, the result. Yet where have the most important innovations that have changed our lives in recent years come from? Google, Facebook, Zipcar, Wikipedia, and Kickstarter were all founded by individuals, not big corporations. Which is not to say that big companies can't do innovation right—but they'd do well to look to start-ups for guidance.

Pivoting often requires charisma, a relationship with the community of people invested in your project: team members, partners, and a devoted audience. Today's most creative individuals see their work as a calling; that belief in their work gives them the energy to move forward and inspire others to join them in what becomes not just a business but a social movement. They cultivate their charisma in order to serve their calling, and so can you.

TOGETHER THESE COMPETENCIES GIVE US a new foundation to build a more vibrant kind of economic system. People with Creative Intelligence are ushering in a new way of doing business, one that's more in keeping with the origins of capitalism than the finance-based model of the last couple of decades. I call it Indie Capitalism because it is free of many of the constraints and notions that we commonly associate with the economy. We can already see the broad outlines of this system as it evolves.

Indie Capitalism tends to be socially, not transactionally, based. Networks, not markets, are the core component. Value is created by making the new, not trading the old. Indie capitalists care about what's meaningful to people, not what their "unmet needs" are. The locus of value in Indie Capitalism shifts toward the local economy and away from the global economy. Globalization remains important, but even the largest multinationals understand the importance of adopting "local" values like generating jobs in the neighborhood, sourcing from nearby farms and factories, and crafting sustainable products and services.

MY GOAL IN DEVELOPING THE concept of Creative Intelligence is to make the practice of creativity routine. I believe it can be an organic, everyday occurrence, not an artificial experience orchestrated by consultants who encourage participants to wear funny hats and write wild ideas on a whiteboard. I'd like to enable you to create easily and often. I'd like to encourage you to become comfortable with your own creative habits and to rediscover how much fun they can be. I want to reintegrate the concept of play with the concept of work and show how the rituals of creativity cross from one dimension to the other.

For some people, building upon their Creative Intelligence might mean taking an edgy photo and sharing it with Instagram. For others, it might mean launching a storefront on Etsy or Amazon. We all have the ability to make things, and while we might not know how to do so just yet, the tools that make creation possible exist and they have never been as inexpensive to access or as easy to master.

Creative Intelligence is about tools, not lightbulbs. It's something we do, not something that happens to us. It's about what happens during those moments of insight, but also *after*; it's the hard work and the collaborations that can help bring your idea out of your mind and into the world.

Creativity can be common and routine, not rare and occasional. It is something that can be evaluated by assessing performance at specific tasks or reviewing a body of work, not measured by standardized tests. Above all, Creative Intelligence is a way of expressing our humanity, our unique power to create, connect, and inspire.

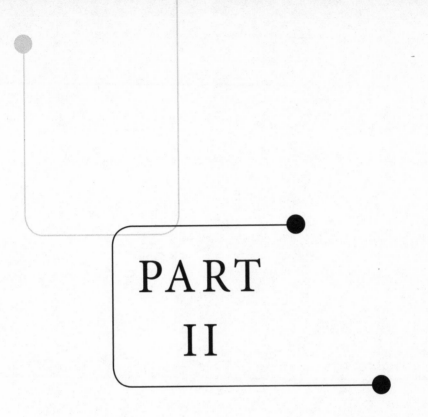

PART II

The Five Competencies of Creative Intelligence

Knowledge Mining

STEVE JOBS WAS A VERY private person, but his commencement speech at Stanford on June 12, 2005, marked one of the few times he opened his heart in public. Jobs, then CEO of both Apple and Pixar, told the story of dropping out of Reed College after just six months because he felt it was too expensive for his working-class parents.

Rather than abandon his education completely, Jobs lingered on campus, going to the classes he really wanted to attend. During that time, he slept on the floor in friends' rooms, returned Coke bottles for the five-cent deposit, and ate his one good meal for the week at a Hare Krishna temple. But despite all that, Jobs's time at Reed was invaluable. "Much of what I stumbled into by following my curiosity and intuition," he said, "turned out to be priceless later on."

One of those classes he sat in on was calligraphy, which he described as "beautiful, historical, artistically subtle in a way that science can't capture." Back then, the practicality of learning calligraphy didn't even occur to him. His goal was not to apply it, simply to learn and enjoy it. "But ten years later," he said, "when we were designing the first Macintosh computer, it all came back to me. And we designed it all into the Mac. It was the first computer with beautiful typography."

At the time, in the moment, Jobs never could have imagined he'd one day connect calligraphy to the design of a computer he'd develop years later. As he put it, "You can't connect the dots looking forward; you can only connect them looking backwards." What's more important, he explained, is that you seek out those dots, and trust that you will connect them somehow in the future. Something, he said, "your gut, destiny, life, karma, whatever," will help you make the connection.

Jobs was talking about a different style of learning and a particular kind of knowledge—neither of which fits squarely into a one-size-fits-all educational system or idea of success. He followed his own path, and the insights he had along the way set him up to make surprising connections down the line.

I'd like to suggest that we don't need to leave it up to fate to make those connections, and that Jobs really didn't think so either. It is certainly true that serendipity affects how and when we connect different ideas to produce new ones, but there are behaviors we can learn to better facilitate these connections down the line. This kind of "Knowledge Mining" can involve studying what came before, as Jobs did with calligraphy classes, or "mashing up" two seemingly unrelated fields or bodies of knowledge to create something new.

In this chapter I introduce several different kinds of Knowledge Mining. "Embodiment" involves becoming aware of knowledge and skills you may not have even recognized as creative and putting them to use in new and surprising ways. "Immersion" is the kind of knowledge we're perhaps most familiar with: throwing yourself fully into a topic or culture that interests you, honing a new skill or area of expertise until it becomes second nature. We don't need to be experts to be knowledge miners; we just need to look around us. Skilled innovators also "mine the past"—looking to the history of their industries and beyond for ideas they might revisit and revolutionize. Like Steve Jobs, they "connect dots"—

following their curiosity, trusting that one day it will help them make surprising connections.

Over time, the more knowledge you mine, the more dots you connect and patterns you create, the better you'll be able to see when something's missing. I call this "donut knowledge." It's a skill that's not as rare as you might think, though we often call it by different names. We'll say we have a "hunch," or feel something "in our gut." People with donut knowledge are said to possess a "sixth sense," or they're called "streetwise."

All of these strategies would be useless, however, without the final one: an ability to understand what people find meaningful. Together, these skills help us build on knowledge we already have and seek out new knowledge. Sometimes Knowledge Mining means going deep into a subject; other times we need look no further than ourselves. It also means having the confidence to follow what excites you even if you're not sure how this new skill or body of knowledge will help you quite yet.

Fortunately, you don't need to wait a decade for your insights to start coming together. Whether consciously or not, you have been accumulating knowledge since birth.

EMBODIMENT

Roommates Adam Lowry, a climate scientist for the Carnegie Institute, and Eric Ryan, a junior ad agency account planner, were both unhappy at their jobs and looking for something else to do.

Lowry was tired of writing research papers on climate change that few people read and which had little impact. Ryan's most recent project, a campaign for Colgate toothpaste, had him wandering the floors of grocery stores looking at what he would later describe as "a sea of sameness."

Lowry, meanwhile, was thinking about how the drab look of most green products projected the message that sustainability was about sacrifice. "You're paying your penance for the Earth that's dying, and you need to save it," he said. "I was continually frustrated that you buy these green products and they cost more, they don't work, they're not fun to use. It's like guilt and absolution, instead of living a positive, healthy lifestyle."

At some point, Ryan told Lowry that even though he knew nothing about cleaning (their apartment actually was pretty dirty), he was thinking of launching a detergent start-up. Lowry jumped at the opportunity and began to devise eco-friendly detergent formulas in his head.

The two used $90,000 that they'd saved and received from family and friends to create a business they called Method. They launched their first product in February 2001, hardly the best time to start a new company, but the young men understood something that Procter & Gamble and other big detergent companies didn't—most members of their generation didn't just think about sustainability; they believed in it. It was part of them. Moreover, unlike boomers who viewed sustainability as a kind of punishment, people their age wanted it to be cool. They wanted products that were stylish and beautiful. And they wanted them at the same price as the old stuff. Lowry and Ryan knew all that because it's what all their friends wanted. And it's what they wanted. "To us," Lowry has said, "'sustainability' and 'green' are just aspects of the quality of our product. They are not a marketing position. . . . I mean everything should be that way."

So they persevered. The big break came in 2004. Ryan, Lowry, and Josh Handy, whom they hired to run product design for their company, deconstructed the leading soap detergent at the time, P&G's Tide brand. It was 90 percent water, according to Handy, which made each package of Tide both heavy and large.

You could save a lot of energy that went into materials and transportation and cut wastes significantly simply by concentrating the formula and shrinking the container size. Wrap it in a stylish package and voilà: a Gen Y–friendly eco-friendly laundry product. When Method persuaded Target to carry its products, the mass market for eco-friendly well-designed products was born.

In 2007, the company did further research and found that 57 percent of people use twice as much laundry detergent as is necessary. So Handy designed a new pump that delivered just the right amount of detergent as well as a simple stylish bottle that weighed less and cost less to produce. Four pumps for a major wash, fifty washes per bottle: less is more.

The Method story is about the changing meaning of sustainability. For previous generations, going green meant giving up comfort, quality, and design. The phrase "limits to growth," the title of the groundbreaking book from 1972, sums up the 1960s and 70s feelings about sustainability—but Lowry and Ryan are members of a generation that doesn't think in terms of limits.

The Method founders built a business model around the values of their generation. "We believe in inside-out branding," Handy has said. "If we don't like it, we feel you won't like it." Method containers, with their colorful and practically voluptuous look, call to the mind the comic books that so many of their peers would have read as kids. So it's not surprising that on the Method website Lowry and Ryan describe themselves as superheroes: "And like every great superhero, they gained their powers after being exposed to toxic ingredients. Cleaning supplies, to be precise. But rather than turning them green or granting them the ability to talk to fish, Eric and Adam's toxic exposure gave them something even better. An idea."

In their desire for a modern, well-designed product that also happened to be good for the environment, the Method guys embodied the values of an entire generation who believed that sustainability is normal, natural, and cool. Maybe very cool. They understood exactly what members of their generation wanted because they wanted it themselves.

CONVENTIONAL WISDOM TELLS US THAT it takes thousands of hours of practice and hard work to become an expert at something. We know that standout athletes like Tiger Woods and Roger Federer trained from the time they were children. We've heard stories about violinist Itzhak Perlman practicing the fiddle when he was three. The idea that we must sacrifice so much to build up a domain of knowledge can lead us to doubt our own abilities to ever be creative. After all, for every Perlman, there are millions of people who remember their violin or piano practice with dread.

What we overlook is the knowledge we all possess by virtue of who we are and which "tribes" we belong to, the kind of knowledge gained not from practice, but from life experience. And yet, the ability to recognize the unspoken aspirations of our own groups and cultures and become a means for fulfilling those dreams is a skill that many creative people share.

You can, like the Method guys, learn to embody the aspirations of your generation or your community, or you can embody the dreams of people on a wider scale. Think Steve Jobs with his ability to see that people wanted an ecosystem of easy-to-use digital tools. This may not have been an aspiration of every member of the boomer generation, but Jobs was, after all, a child of Mountain View. As a teenager, he hung out in the garage of his neighbor Larry Lang, an engineer who got Jobs into the Hewlett-Packard Explorers Club. When he needed parts for a frequency counter he was building for the club, he called HP's CEO Bill Hewlett directly and spoke to him for twenty minutes, according to Walter Isaacson in his biography of Jobs. One could argue that the Apple cofounder's early introduction to electronics, his friendships with other techies like Steve Wozniak, and his growing up in a hot high-tech culture were instrumental in his development as a tinkerer and designer of computers.

Embodiment begins with knowing yourself—who you are, what cultures you belong to, and what you want to create in the world. From the Park Slope mom who builds a babysitting sharing site for her community to the young doctors who embrace social networking in their practices, we all have had experiences that, if mined for their true value, can help us customize our careers and lives. We often don't see this as knowledge because it's something we understand intuitively rather than something we were "taught."

Embodiment is certainly not necessary to creating. But it is a pool of knowledge that we often overlook and underappreciate. And in a moment of hyperconnectivity when ideas are only as good as your ability to implement them faster or better than anyone else, it makes sense to start where you are. You can save yourself a lot of time researching "unmet needs" and a lot of money doing extensive ethnographic research by taking a good look at yourself, the cultures you belong to, and the beliefs and values you embody—perhaps without even realizing it.

When it comes to embodying the values of a generation, a gender, a culture, or even a region, you can't be an immigrant. Sure, you can learn many of the facts, but except for the rare individual who, like a nonnative speaker, gets the accent just perfect, your level of nuance will rarely be as deep as the native's. That doesn't mean that you can't participate in the rituals of a culture or create a piece of art or a product that is embraced by people outside your generation or network, but accept your outsider status and be willing to partner with people whose insider knowledge exceeds your own.

Take Internet culture: If you're in your fifties or sixties, you're every bit as much an immigrant to the new technology as you would be if you moved to France with no knowledge of the language or local customs. You can teach yourself a lot about it, you can immerse yourself in it, but most boomers will never "get" what their kids and grandkids understand about the latest technology and the culture that comes with the technology.

Of course, just as there are some people who master languages with ease, others are adept at navigating Internet culture. The fact that over half of all adult social networking site users are over the age of thirty-five proves that boomers and older Gen X-ers are hardly averse to diving into the new world. But the majority of us will simply not be as skilled at using these new tools as the people who've grown up with computers as their main form of communication and tool for learning. Lest boomers feel alienated by this, lots of my Generation Y students tell me they can't understand their younger brothers and sisters—neither their slang nor their preferred modes of communication. They are already using technology in a different, "cooler" way than their older siblings. The acceleration of change is that fast.

But this shouldn't dissuade us from learning something new. The key is awareness: about your skills, your knowledge, your beliefs—and the gaps in your experience that might make it difficult for you to really "get" a new group or audience right away.

Taking a look at your life to assess what you already embody can provide you with new levels of creative confidence. First, step back, taking a moment to be aware of just what you embody—and what you don't. I recently talked with a woman applying to the fashion school at Parsons. It's one of the top two or three fashion schools in the world and many of us know it through the TV program *Project Runway*, which is often shot on location at Parsons on 37th Street, near Times Square in Manhattan. The woman came from a small town in New Mexico and had watched the show for years. It was her motivation for becoming a fashion designer. But she had kids and a husband and worried whether she could compete against the young kids applying for admission at the fashion school.

What she didn't realize was that she had already developed a key skill in her time as a mother. She made most of her children's clothes and in fact had been making clothes her whole life. While she had traveled to New York to see the exhibit on Alexander McQueen at the Metropolitan Museum of Art in 2011, she didn't know that she shared a skill with the designer known for the "savage beauty" of his clothes. It was a skill McQueen had picked up as a young apprentice cutting and designing men's suits in London, a skill that this woman had learned from her mother, and a skill that most of her young competitors did not possess: the ability to sew a perfect seam.

Knowing what you know is critically important. Because embodied knowledge is tacit, we have to consciously make an effort to understand and value it. Students, for example, are often desperate to get great summer internships at companies and nonprofit organizations to get "experience," network, and possibly get a leg up in being hired when they graduate. But what do they offer managers in return (beyond a willingness to fetch coffee)? What value can they present when they're only eighteen, nineteen, or twenty?

A great deal, as it turns out. Members of Gen Y "get" open-source technology and are comfortable sharing in ways that older people aren't. They have experience building online communities. They grew up in a visual culture and can manipulate images and videos with ease. They are naturally participative and jump into new platforms without thinking.

When you're born into a culture, whether that culture is a

generation that lives on the Web or a community that lives on an island off the coast of France, you're sensitive to the local customs, the rituals, the taboos. When you're an immigrant, everything's new, often exciting. But it's harder to understand what's really going on below the surface. It's more difficult to put things in the proper historical or cultural context. You have to work at it. That's where immersion comes in.

IMMERSION

It would be a fool's errand to attempt to embody the dreams and aspirations of every culture. But not all of us want to remain in the culture we were born into. We want to explore new places, meet new people, and try out new ways of life. For every Mark Zuckerberg who built a new company upon the values of his generation, there are creative individuals who left their worlds behind in order to gain new skills in new areas. Howard Schultz was not born in an Italian coffee culture but traveled to Italy to learn what made it so special and then brought it back to the United States and built the coffee experience into what we know it to be today. For every Method, there are stories of companies that relied more on research than personal experience to connect with customers. For those of us who don't already embody the values of a group we're trying to reach, we can team up with those who do, or immerse ourselves in learning about that group's beliefs and habits.

As one of China's more innovative privately owned companies, Lenovo has had a history of competing strongly against foreign brands in China. The generation that came of age after the disastrous Cultural Revolution was happy to buy what was offered to them and Beijing-based Lenovo provided them with great computers. But as one century turned the corner into the next, China was changing radically, and a new generation of Chinese consumers—one born into prosperity—proved to be much more discerning. Lenovo's major competitors—HP, Dell, and IBM—were beginning to take market share away by offering lower prices and, because of the appeal of Western products, higher prestige.

The company's leaders had grown up alongside an older generation of Chinese consumer, and so understood their desires to work hard, advance their families, and give their children a better education and life than they had. But like many successful start-up corporations that grew up within a particular generation, Lenovo didn't have the internal competency to understand a new consumer base.

So in the early 2000s, Lenovo turned to Ziba Design out of Portland, Oregon, one of the world's top innovation and design consultancies. Like IDEO, Continuum, Smart, and other firms that began as industrial designers and focused on making consumer products, Ziba had by then evolved into a broader consultancy able to do design research and strategy for marketing and branding.

Ziba knows how to get deep inside a culture. Ziba employees, founder Sohrab Vossoughi told me, represent eighteen nationalities, speaking twenty-five languages. The Portland office has a staff of more than 100 people with more than sixty-two areas of expertise, including Color Specialists, Environmental Designers, Information Architects, Anthropologists, and Cognitive Scientists. This diversity accelerates the firm's ability to know—to synthesize findings and recognize patterns (or the absence thereof) faster than any one expert could.

Lenovo wanted to compete not simply on price or prestige but on meaning and value, so Ziba went to work identifying those values. Even before they left for China, Ziba's team of social scientists, design researchers, and product designers surrounded themselves in a distinct project "war room" filled with Chinese billboards where they listened to Chinese rock and classical and traditional music throughout the day. Chinese exchange students were brought in to interpret lifestyle and technology magazines, especially the advertisements. Objects young Chinese used every day—wallets, cell phones, cigarette lighters—were collected and analyzed in terms of the choice of colors and textures and finish.

Once in China, the team split into two groups, and both spent four weeks immersed in three different regions. Design anthropologists, design strategists, and industrial designers talked on cell phones as they commuted on bicycles with Beijing workers. They ate from street carts and dined on pig brain and pigeon in large banquet halls. They walked the ancient Hutong alleyways and sang late at night in karaoke bars. Each team rode buses and trains, wrote text messages in nightclubs, and used notebook PCs in Starbucks. Visual inspiration was drawn from fashion boutiques and electronics stores, from traditional gardens and modern architecture. Ziba's people even spent time going through closets in young people's apartments to note fashion tastes.

Ziba also enlisted the help of Chinese consumers. Volunteers were given a camera, a glue stick, and two poster boards and asked to photo-document one work and one leisure day, giving special attention to moments when they integrated technology into their routine. The volunteers then created visual time lines, which gave the researchers a glimpse into daily behaviors and emotions.

When the team returned to the war room in Portland, they distilled the visual worksheets, photographs, and observations from each interview into a single Ethnography Inspiration Sheet.

Ziba then used the Inspiration Sheets to identify the aspirations and behaviors of distinct clusters of Chinese tech consumers. "We called these clusters 'technology tribes,' said Vossoughi. "We found five new tribes in China."

The tribes included Social Butterflies, Relationship Builders, Upward Maximizers, Deep Immersers, and Conspicuous Collectors, each of which possessed vastly different desires, ranging from the desire to connect with a broad social network (Social Butterflies) to the desire to seek escape through fantasy and immersion (Deep Immersers). These profiles gave Ziba a way to work with Lenovo to gauge the size of each market segment.

Ziba used improv acting techniques to understand how, for example, a Social Butterfly would use a cell phone compared with how a Deep Immerser or a Relationship Builder would do so. The project for Lenovo, called Search for the Soul, led to a new understanding of who Lenovo's target consumers ought to be and laid the groundwork to create product-line strategies for Lenovo's desktop, notebook, and cellular platforms.

In the end, Lenovo used the understanding that came with immersion to make three new products—a desktop PC for Deep Immersers, a notebook/tablet PC for Relationship Builders, and a cell phone for Upward Maximizers—that addressed the unique needs of these new customer tribes.

The knowledge gained allowed Lenovo to push back its foreign rivals and increase market share in the Chinese market. In 2012, Lenovo had about 30 percent of the Chinese PC market, ahead of Acer, Dell, HP, and Asustek. ZIBA's immersion in the Chinese market was also important in Lenovo's decision to purchase IBM's ThinkPad division in 2005, bringing it neck and neck with HP in the world PC market.

Immersion is the kind of Knowledge Mining that we are most familiar with. As students, we immersed ourselves in history and literature and science in order to understand them. But rarely do we view immersion as a creative competence. We don't always see the knowledge we've gained as the result of deep study or practice as a resource that can be mined to generate creativity. And we should.

The story of Ziba's work in China illustrates that immersion can be instrumental. This can take time. In order to understand Chinese youth culture, Ziba sent twelve researchers around China for three months, spending a combined 205 days, or nearly five thousand hours, gathering and analyzing data and formulating personas to represent their findings. A similar project Ziba did for Li Ning, China's rival competitor to Nike, took nearly nine thousand hours. For a nuanced understanding of complex and rich subjects such as national and generational cultures, this kind of deep immersion is very important.

Immersive knowledge can't be shallow, but it needn't be quite so deep. Sometimes taking a single class or doing a little research about a topic you're interested in is all it takes. Alexander McQueen might not have known that studying his Scottish roots would shape his breakthrough runway show in London—the controversial "Highland Rape" show that would establish his position and fame until his suicide in 2010. Steve Jobs had no idea that when he wandered into that calligraphy class he was about to learn something so important that it would shape the design of the user interface for Apple computers and, eventually, all digital products.

The class had an enormous impact on his life—and all our lives. But it didn't take a lot of time. There is a critical distinction between knowledge that makes you an expert and knowledge that allows you to be creative. One takes a very long time. The other doesn't necessarily have to take so long. We can all immerse ourselves long enough to build up a knowledge base that we can mine for creative purposes.

If creativity is your goal, it may be a better strategy to spend less time practicing to become the best in a particular field and more time learning many kinds of knowledge. Daniel Pink, author of *A Whole New Mind* and *Drive*, told me in a webcast that the fastest growing major in the United States today is the double major (one popular choice combines computer science with art and design). Students are intuitively expanding the range of their knowledge and as a result boosting their Creative Intelligence.

Nearly all of us love to learn new things. We study, we practice, and we spend time gathering information. We often don't see the role this learning plays in boosting our creativity. And yet almost everything that we learn may have value in innovating. That class you took, the book you just read, the salon you went to, or even the eighteen-minute TED video you watched can all be jumping-off points for creating something new. Following your own curiosity and immersing yourself in what interests you, not what you think you need to learn, is often what will prepare you for work in an entirely different job or endeavor.

We often think that following our curiosity is a waste of time. We're told to study hard, get the right answers to the questions, focus. Budgets for music and art and other "extraneous" courses are the first to get slashed; frequently such programs are cut altogether. Yet by immersing ourselves in a wide variety of subjects, we gather more dots of knowledge that are so essential to creating.

Immersion is an open-ended and ultimately liberating activity. We can all choose to learn whatever interests us. We can build as many dots of knowledge as time and effort allow.

CONNECTING DOTS

In 1990, MIT roboticists Rod Brooks, Colin Angle, and Helen Greiner launched a start-up called IS Robotics. Their first area of focus was space exploration vehicles; they worked on the rovers for NASA that led to the Sojourner exploration of Mars in 1997. According to their website, they soon began developing in other areas, receiving government grants that led to creations like the Raptor, a robotic dinosaur; DARTS, a robot designed to move like a fish; and Ariel, a mine-hunting robot with an organic, biological shape. The company also worked with the toy-maker Hasbro on My Real Baby, a robotic doll that was ultimately discontinued, but not before the team learned how to design for low-cost manufacturing.

But the research on space exploration and toys didn't exactly make a difference in people's everyday lives, and one of the founders' major goals was to make practical robots a reality.

In 1997, the company won a government grant to begin work on robots that could search for victims of disasters. The resulting robots would be instrumental in the search for survivors in the World Trade Center and the Pentagon following the attacks of September 11, 2001, and later funding from DARPA led to the creation of the PackBots, used to search for the Taliban in Afghanistan and to disarm bombs in Iraq. PackBots would also be instrumental in removing debris and measuring radiation at Japan's Fukushima Daiichi power plant following the earthquake and tsunami.

The same year it won the grant to begin work on the robots, the company landed its biggest contract yet: one from SC Johnson

Wax to work on the AutoCleaner, a huge automated industrial floor cleaner. After the project, two engineers from the design team connected their new knowledge about cleaning with what they had learned about low-cost manufacturing from their work with Hasbro. The company counts "building on what we know" as a major aspect of its product development strategy, and because its robots are component-based, it can easily take an approach used in one area and apply that to a new technology, rather than build from scratch.

It took a long five years to prototype what became the Roomba, the world's first vacuum cleaning robot. (The company changed its name to iRobot in 2000 after merging with Real World Interface, a New Hampshire–based robotics company.) Since its launch in September 2002, more than 5 million Roombas have been sold—there's even a smallish cult of Roomba "hackers" who manipulate features and paint the robots to give them personality.

The story of the Roomba is an example of Knowledge Mining at its best. The company hadn't really created anything "new"—the technology of the Roomba was utilized in other forms in their many other projects. Nor was the robot the result of a flash of insight. Instead, it took time and hard work—a full decade—for the engineers to connect dots that no one else had dreamed of connecting in order to create the Roomba and make robots, long the love of science fiction writers, a part of everyday life.

Knowledge Mining doesn't require spitballing hundreds of ideas, a common brainstorming technique. The first step toward building bridges that connect seemingly unrelated islands of knowledge—a way of thinking that's been called "lateral," "relational," or "horizontal"—is often as simple as looking at what you or others have done in the past and thinking about how you might expand on that to create something entirely new. But you do need a strategy.

Early advocates of this approach at Procter & Gamble abandoned their traditional R&D process where all the research took place inside the company for what they called Connect + Develop in 2000. The name speaks for itself. Connect + Develop has generated billions for P&G since the company began requiring all divisions to open themselves up to outside sources of information and inspiration.

P&G may be seen as a mass-market consumer company, but its core culture revolves around chemistry. A brilliant cadre of chemists has, over the decades, developed formulas for cleansers and other household products including Tide, Pampers, and Crest, which are then packaged, branded, and sold around the world by an army of marketers and salespeople. Virtually all of P&G's products came from its in-house chemists.

In 2000, newly appointed CEO A. G. Lafley broke open this culture and encouraged P&G's division heads to start looking outside their divisions and even outside the company for new ideas. P&G set up a series of networks: top P&G officers from around the world were teamed up with technology entrepreneurs to mine existing science journals, visit conferences, and meet scientists whose work might be useful to future projects. They have worked with the NineSigma network to facilitate communication with other science-based companies and universities, and have also teamed up with InnoCentive, a network that links companies with more than 260,000 "problem solvers" who are paid to solve challenges posed to them.

The new networks opened up P&G's silos to product concepts and synergies that managers and scientists could not have come up with on their own. One result was the Crest Spinbrush, which P&G purchased from a group of entrepreneurs led by

John Osher in 2001. Others were new combinations of existing P&G brands, such as Mr. Clean Magic Eraser and Olay Regenerist. At least a third of the innovative new products at P&G now come from "Connect + Develop."

While the idea of building bridges in your own organization or drawing from your own experiences to innovate something new might *sound* appealing, anyone who's attempted doing so will tell you it's not always that easy. With a countless number of choices to consider, where do you begin? In a universe full of infinite dots of knowledge, which dot is the right one?

Casting for Ideas

It's easy to get carried away in the hunt for ideas; if you chase everything shiny and fast, you risk forgetting what you're seeking in the first place. In a way, it's like fishing: It makes sense to get your bearings before you start casting. As any person who fishes can tell you, when you're casting, there is no way to foretell exactly what you will catch; the key is being open to inspiration from unexpected places. Scientists at the Taiwanese-based Industrial Technology Research Institute discovered this during their search for the most recent holy grail of computer technology: a flexible screen for our iPads, smartphones, PCs, and TVs—one that you can roll up and out, like a scroll.

Scientists at the Institute had been working on a way to get that flexibility, which involved placing transistors onto a flexible substrate that is placed over glass. Transistors are transferred to the flexible substrate and it is then lifted off the glass. The problem is stickiness. It is very hard to lift the flexible substrate from the glass.

According to the *Wall Street Journal*, ITRI division directors Cheng-Chung Lee and Tzong-Ming Lee found a possible solution after being inspired by one of Taiwan's favorite dishes: pancakes. As any chef knows, the high temperature used to heat the oil makes it very easy to peel the flexible pancake off the pan. Back in the lab, the scientists were able to replicate the pancake process by placing a layer of nonadhesive material between the substrate and the glass. A flexible display for an e-book reader should be on the market soon.

Casting wide can land you in the strangest of places, but having some kind of anchor—a puzzle you're trying to solve, a product or process you're trying to improve, or some knowledge about a new technology or a group you're trying to reach—can help keep you from floating too far off-course.

We don't always know what connections will work best, what synthesis of two ideas will be the most effective. That's why it's important to keep an open mind about what your casting may bring back. You may have to relearn the joy of surprise, as it is often in the surprise that we find solutions. When James Dyson went casting about for a new technology for vacuum cleaners unlike the standard bag filters, he found it in sawmills, which traditionally use industrial cyclone fans to suck up the sawdust. That spinning cyclone method of picking up dirt became the heart of his Cyclone brand of vacuum cleaners, a successful cast if ever there was one.

You're not born with a great ability to connect dots. You learn it. Some of us learn it in school, some at jobs, others in life. It's not a difficult competence, but it is a deliberate one. The anxiety many of us feel about creativity often stems from a belief that we need to create something from nothing. We don't. The scientists at ITRI and James Dyson began with something they already had—an area of expertise, a skill, a technology they were hoping to update—and went casting for ideas in both new and familiar places. What these innovators share is an ability to harness the serendipity of life to fashion something wholly original.

Mining the Past

When I visited him in Toronto in 2011, Bill Buxton had just finished working on a birch bark canoe using traditional Cree Indian instruments. He was getting ready to follow part of the old fur-trade water route on the Churchill and upper Sturgeon-Weir Rivers to watch the northern lights, but he had the time to launch into an important discussion on context, connectivity, and creativity.

Buxton is a true Renaissance man. Not only is he able to make a Cree birch bark canoe; he has been Chief Scientist at software graphics companies Alias/Wavefront and SGI and a professor of computer science at the University of Toronto, and he is now Principal Researcher at Microsoft Research. Buxton began his career as a composer, musical instrument designer, and performer; after receiving his Bachelor of Music degree from Queens University, Canada, he began designing his own digital instruments. He then got a computer degree and headed over to Xerox PARC, that remarkable innovation arm of the Xerox Corporation that first developed the computer mouse, menus, windows, and graphical user interfaces, all elements of digital devices that we take for granted today.

The day we met, Buxton and I spoke about one of my favorite topics, Apple, a company that Buxton greatly admires. No CEOs, managers, or designers in the corporate world argue with Buxton that Apple is one of the most creative companies in the world and that senior vice president of Industrial Design Jonathan Ive is perhaps one the most successful industrial designers of our time. Yet, Buxton would point out, few of them actually know how Apple and Ive "do" their creativity. Because of this, most companies, particularly in consumer electronics, do design innovation wrong.

People need to do their homework, Buxton argues. They need to take the time to study the history of their field, the successes and failures of the early years. While many CEOs and managers want to be like Jobs and Apple, they haven't taken the time to study how they work and where their core ideas come from. Young techies can often "listen to Jimmy Page, Eric Clapton, or Keith Richards—some of the most influential guitar players of the past fifty years—and know (a) that they are riffing on what the great blues artists did in the past, and (b) recognize the riffs and where they came from," said Buxton. But they can't do the same for their own companies or industries. "The prevailing myth amongst those very same wannabes is the notion that creativity and invention is of the instant flash of invention lightbulb variety," said Buxton.

Buxton can cite four examples of how Ive, Jobs, and Apple riffed off the past to create Apple's successes. Take, for example, the original iPod. "The iPod quotes the 1958 Braun T3 transistor radio, designed by Dieter Rams," he said, referring to the great German designer for the consumer electronics company Braun. It's very Bauhaus, with clean lines, simple functionality, just like the iPod. Buxton even sees the historical connection between the launch strategy of the iPod mini and the marketing of the 1928 Kodak Vest Pocket Camera. Teague, one of the first design consultancies in the United States, launched the Kodak in five colors, "the same five colors later used by Apple," said Buxton. Not a coincidence.

Strange as it may seem, the most advanced technology in products is often decades old. Buxton's interest in the history of technology is evident in his collection of interactive gadgets— perhaps the biggest in the world. He has a first-generation Etch

A Sketch, an electronic drum set, watches, keyboards, a Nintendo Power glove, and hundreds more. He's donating the collection to the Cooper-Hewitt National Design Museum, formerly run by his old friend, the late Bill Moggridge, who was cofounder of IDEO and a world-famous interaction designer himself.

Buxton's favorite story about mining the past for inspiration involves a gadget in his collection called "Simon," a 1993 IBM/Bell South smartphone, the first-ever of its kind. The only physical buttons on the Simon were the on/off and volume buttons. All the rest of the Simon's functionality was accessed through a touch screen, one that covered the entire face of the phone. Most of us don't remember Simon nor do we realize that touch screen technology existed two decades ago, but it was known in Silicon Valley when Ive began to work on Apple's ill-fated Newton, one of the first PDAs. Buxton believes Simon to be a likely inspiration for both the Newton and the iPhone.

Artists, dancers, and writers have long known the importance of mining the past. Vincent van Gogh was inspired by many artists, but perhaps none more so than Jean-François Millet. "I put the black and white by Delacroix or Millet . . . in front of me as a subject," he said, describing his process of "copying" twenty-one of Millet's works to his brother Theo. "And then I improvise colour on it, not, you understand, altogether by myself, but searching for memories of their pictures; but the memory, the vague consonance of colours which are at least right in feeling, that is my own interpretation." Van Gogh believed that his process was more akin to "translating them into another language than copying them."

Bob Dylan looked to Woody Guthrie as a source of inspiration. According to Caspar Llewellyn Smith, writing in the *Guardian*, "Dylan started mimicking his hero's speech patterns and even told the crowd at the Cafe Wha? when he arrived in

New York for the first time the following January: 'I been travel-lin' around the country, followin' in Woody Guthrie's footsteps.'"

More recently, Lady Gaga's music, style, and videos reference so much of Madonna's body of work that a number of articles have been written about whether what Gaga has done is homage or flat-out copying. When asked in an interview with ABC News's Cynthia McFadden what she thought of Gaga's song "Born This Way," which shares a chord progression with the eighties classic "Express Yourself," Madonna reflected, "It feels reductive."

"Is that good?' asked McFadden.

"Look it up," said the Queen of Pop, smiling devilishly before reaching for her mug and taking a sip.

While perhaps more rare than in the world of art and music, there are those in the business world who've learned to mine the past. Elon Musk, the founder of SpaceX, the first private company to send cargo to the International Space Station, has a replica of the Saturn V, the powerful rocket that sent twenty-four astronauts to the moon as part of the Apollo program in the sixties and seventies, on his desk. He no doubt has looked to the Saturn V as inspiration for the development of his Falcon rockets as he seeks to further commercialize space. And, of course, the auto companies are constantly looking back to models that we remember so fondly. BMW bought and revived the British MINI Cooper, a brilliant 1959 design for a new small car that Alec Issigonis developed and Jack Cooper improved in 1961. It is once again one of the world's top-selling city cars.

Not only does building up a body of knowledge help you see where things came from and where to go next, it also helps you see what's missing.

DONUT KNOWLEDGE

I was flying home from a finance conference in Europe in the late eighties, sleeping in an aisle seat, when liquid hit my face. Startled, I jumped up, put my glasses on, and looked around, afraid that the plane was in trouble. After a bit, I could see a small figure far down the plane coming back toward me. As the figure moved, passengers jumped up, startled, feeling their faces as I did. Finally, the figure came right up to me. It was a little girl. She was holding something I hadn't seen before, a donut-shaped milk bottle with a hole in it. She was able to hold onto one side with her little hand and squirt the adults in the aisle seats and then squirt milk into her own mouth. She was laughing, happy. She could feed herself—and torture adults while doing it. This was empowerment. When I got back home, I wrote my first article for *BusinessWeek* on design.

It turned out that the girl's parents were Midwestern product designers; they'd tried conventional bottles, but they were all too big for their daughter to hold. Taking away the middle of the bottle and making it donut-shaped meant that small children could easily hold it. I don't believe the designers envisioned people getting squirted on planes as a result, but I do think they understood something fundamental to knowledge. Sometimes what's not there is more important than what is.

Paul Polak has spent his life cultivating a kind of wisdom I call "donut knowledge"—the ability to see what isn't there." Since he began his career as a social entrepreneur in the early 1980s Polak has talked to thousands of villagers in Africa and Asia. He's seen what happens when philanthropists donate money to developing areas—immediate joy followed by disappointment when changes prove only incremental and quality of life remains the same. Polak has also heard a lot about the problems of the developing world—

lack of food, water, hospitals—only to learn that, most often, the solution lies in exploring what isn't being discussed.

When Polak was in Bangladesh in the mid-1980s with his first organization, International Development Enterprises (IDE), he heard a lot about the need for more water for irrigation. Most villages had wells, but it was very difficult to bring it up bucket by bucket by hand. Polak looked for a simple way to tap the well water and found it in a new type of treadle pump powered by leg muscle and the weight of your body that cost around $25.

But after talking to a number of villagers about irrigation and pumps, Polak also learned what was not being discussed when the subject of water came up: income. Sure, people really needed more water for their crops. But the real challenge was to create more income for people in the village so they could climb out of poverty. Polak insisted that all the treadles be made in local factories, rather than in China, so the money paid for them circulated back into the neighboring communities. There are now eighty-four factories making treadles in Bangladesh, with local villagers' family members bringing home wages, while the farmers are boosting their income by increasing their crop yields. Some 1.5 million treadles have been sold in Bangladesh since 1985.

Decades later, Polak found himself working in a different country but tasked with a similar problem. There are 325 million people living in eastern India in states like Orissa, and 80 percent of them don't have access to safe drinking water.

We tend to think that the problems of poor countries are due to scarcity: There isn't enough water to go around. But traveling through eastern India, Polak saw something else. He came to understand two critical realities beneath the conventional wisdom. First, in the state of Orissa, there is plenty of water available in villages, but it is not clean. Human fertilizer in the fields often gets into the water supply. So while the villages have pumps making water available for people (almost always women) to come and carry it back home, it usually contains fecal matter. Diarrhea and other chronic ailments are common.

Second, water reflects political power. Members of the higher castes control access to water where it's available; in one instance, according to Polak, when someone from the "Untouchable" caste touched a faucet in a village, members of the higher caste insisted that the tank be drained, "purified," and refilled. Where water is scarce, higher caste farmers call upon political connections to build dams that funnel clean water to their villages.

Polak's ability to see the hole in the conventional wisdom led to a different kind of solution. The system was rigged and irrigation was expensive, yes, but purification was cheap. He set up a private company, Spring Health, and with the help of Indian design consultancy Idiom designed a business plan: Spring Health pays for a $100 concrete water tank to be set up next to one of the two local kiosks called "kirana shops" that exist in every village. The tank is filled with contaminated local well water and cleaned with chlorine. The shopkeepers then sell the clean water for four cents per ten liters—a day's worth of water. If you wanted home delivery, as many did, it would cost you five cents. Idiom then designed a local transportation system to deliver clean water to families outside the village and designed a new ten-liter plastic jug and a bicycle "saddle" that could hold six liter-size water containers. For a family living within three kilometers, clean water can be delivered for eight cents a day.

In the first six months of operation, in 2012, village medical expenses for diarrhea and other waterborne diseases dropped dramatically. New income was generated for shopkeepers and new water-delivery jobs were created for villagers.

According to Polak, "Spring Health will generate a cornucopia of jobs in the villages. . . . By the end of the first year, if we are successful, we will be partners with six hundred small ki-rana shops in villages, whose livelihoods and status in the village will increase. They in turn will hire bicycle delivery and hub and spoke rickshaw delivery people from the villages." With millions of kirana shops in India, the potential to replicate this model throughout the country is huge.

The Acumen Fund, founded by Jacqueline Novogratz, is now investing in Spring Health to scale it. Polak hopes it can reach 5 million villagers in three years and 100 million in ten. Many, if not most, of the beneficiaries will be Untouchables handling their own clean water, some for the first time in their lives. Po-lak thinks Spring Health could be the first billion-dollar business specifically designed for the demographic C. K. Prahalad dubbed the "bottom of the pyramid."

None of this would have been possible without Polak's deep domain expertise developed over decades of work in the field.

So how do you gain donut knowledge? The simple answer is: time. The more you know the pattern, the better prepared you are to see where it breaks. That can mean leaving the familiar behind, going to places—either physical or intellectual—that are unfamiliar, and sometimes uncomfortable.

While donut knowledge naturally increases with time, there are some strategies you can adopt now to improve your ability to see what isn't there. I've picked them up from my years of birding, but they've been just as useful in other areas of my life.

Know the Pattern, Watch for the Breaks

As a birder, I've learned to look and listen for what shouldn't be there. What's unusual. What goes against popular wisdom. It involves a certain amount of domain expertise—I'm certainly a better birder now than I was when I began fifteen years ago. But even if you've yet to amass experience in a particular field, you can still improve your chances of spotting the surprises you may not be expecting.

For birders, it can mean going to strange and sometimes unsavory places. When I was in Singapore for a design conference, I went birding at a municipal waste treatment facility and found a number of birds—including one black swan. It was a rarity in Singapore and a good find. I was surprised, but not shocked. I was, after all, looking for what was not supposed to be there.

Just as good detectives are trained to hear the dog that did *not* bark—so too are good scientists trained to look, and listen, for what's not there. In 2012, Jeremy Feinberg, a Rutgers doctoral candidate, was doing fieldwork in the marshes and ponds surrounding New York City when he noticed something strange. He was researching the decline of leopard frogs in the area when he noticed that the croaks of the frogs didn't sound like any he'd heard before.

According to CNN, Feinberg knew that the frog's croak was "peculiar," that its calls sounded "weird." He also knew what the croaks lacked—the "long snore" and the "rapid chuckle" leopard frogs are known for. Feinberg had a hunch that this frog didn't fit the pattern. "When I first heard these frogs calling, it was so different, I knew something was very off." Very off.

As we now know, something *was* off. DNA testing showed that Feinberg identified a distinct, new species of leopard frog with a range that encompassed New York City. Leslie Rissler, an associate professor curator of herpetology at the University of Alabama, said the discovery was "extremely rare."

The new species has not yet been named. Feinberg first identified it in the bogs of Staten Island, but it's now been found in two other states. "I've given it lots of thought," he told the *New*

York Times. "Part of me has always wanted to call these New York leopard frogs, but I think people in New Jersey and Connecticut will protest. I have to balance the politics with the naming." I think a better solution would be to name it after the discoverer, the scientist who knew the pattern well enough to hear the break. The Feinberg Frog.

Bird the Birders

Another donut knowledge strategy is to "bird the birders." When I go into Central Park during migration or travel to Central America or the Amazon, I surround myself with better birders. Put many good birders together and the chances of finding a rarity like the red-footed falcon go up significantly.

Whatever your interests are, surrounding yourself with people who know how to look for the odd duck tends to increase your own chances of finding the rarity. You can learn from the experts, see how they operate, and, sometimes, simply tag along.

Universities have long been hubs of creative activity— one reason why Stanford produces so many entrepreneurs and founders of start-up companies is the dense network of people, places, and events that focus on what's next, what's different, what's *not* there—but now there are countless alternatives that can help you connect with experienced innovators from a wide variety of fields. Seek out the right conferences—for every TED or Davos, there are scores of smaller upstart festivals that might feature the thinkers and creators we'll all be talking about in a few years—and go to those weekly evening events thrown by people in your field. Even the expensive and exclusive TED conference has expanded to include hundreds of independently organized TEDx events.

Much has been written about serendipity and the creative congestion of cities—but cities, like universities, need to be more than just crowded. The world is full of huge, immensely crowded cities that don't necessarily generate creativity. Take Singapore, which, for all its newfound prosperity, has a reputation for being boring. Talk to the artists and students from Singapore who have studied in Europe and the United States and they'll say that conformity and mass consumerism define their culture. People almost always buy established global brands like Prada and Gucci rather than support homegrown designers. The government is making a huge effort to promote creativity in Singapore, and it may yet succeed.

But compare Singapore with New York in 2013 and the differences are vast. Over the past fifteen years in New York City, vast networks of entrepreneurs, incubators, venture capitalists, universities, media companies, and artists have arisen to generate a new wave of creativity and entrepreneurialism. For the first time, New York rivals Silicon Valley in start-ups, concentrating more on content and culture than technology. Dial the clock back to the seventies, and you'd have witnessed a different kind of creative congestion—an art scene was burgeoning in SoHo; rappers, break dancers, and beat boxers from all the boroughs were creating a new kind of music and culture—but it wouldn't necessarily have been the ideal place for someone looking to start up a new media company that required a thriving network of graphic designers and social media experts. You need to pick your congestion, your cities, carefully.

DREAMS, NOT NEEDS

When it comes to international sports, India is often outclassed by other nations. But recently, India's female wrestlers have brought home the gold. In the 2010 Commonwealth Games, they won three gold medals. In the 2011 Commonwealth Wrestling

Championship in Melbourne, they took home five golds. It's a remarkable achievement in a country that provides little support for women's sports—or women's rights.

Wrestlers of both genders have had to train without much funding, with state support coming in the form of government jobs and cash rewards upon the winning of medals. There are no programs like Title IX to promote women's sports in schools— and no real government support for women's sports in general. Though wrestling has had a long tradition in India, it has always been a male-dominated sport, despite recent standout performances by women.

But Usha Sharma, a police officer and the wrestling coach who trained several medal winners, dreams of changing that.

Sharma's wrestling academy is based in Haryana, the Indian state known for two things. It is one of the country's most prosperous states. And it is the state with the highest ratio of female feticide. Sons are preferred over daughters in most of India, and the spread of ultrasound technology in prosperous states such as Haryana makes it easier to identify and abort female fetuses, though this practice has been illegal in India since 1971. Haryana has 830 girls per 1,000 boys, compared with the normal sex ratio for children from birth to age six of 952 girls per 1,000 boys. Such statistics are subject to social factors that influence every aspect of the reporting, but they help to guide our understanding of the complex variables at play.

Sharma dreams of building a rural women's sports league, sponsored by local companies, the state or national government, or perhaps all of them. "I have always wanted to do something for women, especially for rural women, who are given few opportunities," said Sharma. "My efforts are focused toward seeing women make their mark in this game, win medals for the country, and make their name."

Social entrepreneurs and philanthropists scour Africa and Asia trying to help the "needy," but Sonia Manchanda, a co-founder of Idiom, India's top business model innovation and design consultancy, believes that asking people about their dreams is a much more powerful strategy. Ask people about needs, she said, and they will give you a long list, one that varies from time of day to day of the week. Ask people about their dreams and they'll give you just one answer, maybe two. It's not a list but a revealing look into what is truly meaningful in their lives.

In January of 2011, Manchanda and Idiom launched Dream:IN, a program dedicated to empowering people all over the country to achieve their dreams. Manchanda brought in 101 students from design, art, and business schools and institutions across India, trained them in the skills of interviewing and filming, and sent them in teams to talk to people in villages and cities across the country.

The students heard many different dreams but identified only about a dozen themes. "More education" was probably the most frequent response, followed by "more rights for women," as well as "help starting or expanding a business." Another dream shared by many of the interviewees was more support for soccer, wrestling, and other popular sports (cricket, despite its popularity, is a sport enjoyed mostly by the rich and middle class).

I joined the process when Manchanda gathered a number of people to Bangalore, Idiom's home base, to strategize different ways to make those dreams come true. Our team came up with a plan to utilize the Internet connection available at many kiosks that dot the streets so people could plug in, free, wherever they were. Private schools and universities could charge a small fee for online courses or the Internet carriers could pay the schools directly because they would receive higher revenues from people using their service. Government and corporate sponsors also could possibly play a role in transforming cyber-cafés into edu-cafés. We also talked of the possibility of delivering education via cell phones, despite the tiny screen size.

A year and a half later, Sharma still awaited funding for her girls' wrestling academy. The project of turning cyber-cafés into mini-schools still remains just a business concept. But Manchanda and Idiom are in the process of building an independent Dream:IN network of venture capitalists, funds, and nonprofits to invest in and fulfill the dreams of Sharma and others interviewed in Dream:IN. They already have $50,000 from Gray Ghost Ventures, a pioneer in microfinance and entrepreneurial investment in low-income countries.

It's not surprising that the Dream:IN model was developed in India—a country where more than 65 percent of a population of a billion people are under thirty-five. Young people, with their lives ahead of them, are full of dreams about the future. But the idea behind the organization is universal. Manchanda's observations about the focus on needs being too narrow is hardly an Indian problem.

For decades, innovation consultancies, brand strategists, and nonprofits alike have spent billions trying to figure out what people need. American and European innovation consultancies offer new strategies for discovering unmet needs. Conventional MBA programs teach managers to get to know the needs of consumers so they might provide products and services to meet them. In the high-tech sector, engineers tried to expand the functionalities of existing products, adding a button here or a click-through feature there for our every need.

We frame commerce and society—even our relationships—in terms of needs, but that kind of framework is limited. People are much more complicated than a list of needs; we need food and housing and shelter, sure, but what makes humanity unique are

the dreams and longings that are much deeper and more complicated than mere necessity. No one *needs* an iPhone or a Zipcar,
and yet these products have become as meaningful to us as the
homes we live in or the food we eat.

Mining for knowledge isn't simply about sifting for data. It
also involves an understanding of what people find meaningful.
Keith Richards and Mick Jagger connected rhythm and blues and
jazz and brought their music to restless kids yearning for change.
Method founders Alex Lowry and Eric Ryan brought together
cool design and sustainability to create a green product that didn't
require sacrifice.

Of course, what's meaningful changes over time and varies
across different cultures and generations. Facebook is meaningful
for twentysomethings and those of us who are a little older and
see the value of this kind of social sharing, but the site will be
less significant to people who believe sharing means having their
grandchildren visit and communicating with their friends over
the telephone.

What are the practical steps that people can take to create
more meaningful things? What should corporations and organizations do to generate more meaningful offerings for their customers and their clients?

Starting with a broad liberal arts degree and continuing your
learning throughout life may be the most important step in gaining the empathy and the context necessary to determine what is
truly meaningful to people. In eras of economic and social stability, specialization makes sense as you seek the best niche for your
talents. You slot yourself into existing organizations and jobs. But
in times of turmoil, when industries and careers are constantly
changing, you need the foundation for building your own path.
Today, the business degree is the most common undergraduate
degree in America, and yet many of the aspects of traditional cor-

porate management—production, accounting, IT, even human resources—are being outsourced to India and China and Latin America. In most business programs, people are learning the specialized skills of the last century. What they need instead is a more general understanding of the world—of its history, culture, art, and science. These are the areas that inspire us to create.

In a talk given to a freshman class at Stanford, writer Bill Deresiewicz cautioned students against getting so caught up in an area of specialty that they lose sight of their ability to make their own paths in lives. "The problem with specialization," he said, "is that it narrows your attention to the point where all you know about and all you want to know about, and, indeed, all you can know about, is your specialty. The problem with specialization is that it makes you into a specialist. It cuts you off, not only from everything else in the world, but also from everything else in yourself."

What we need, perhaps more than ever, are individuals and organizations who are not cut off from but rather intimately connected to the world around them. I'm not talking about superficial connection to thousands of Facebook "friends" you've never met, but connections born from a deep interest and engagement with different people and cultures. It's this kind of engagement that will help us come up with ideas that fulfill our wildest dreams.

KNOWLEDGE MINING: BUILDING A
FOUNDATION FOR CREATIVITY

I recently had lunch with someone from a consultancy widely known for teaching big businesses how to be more creative. During our talk, she admitted that when younger people would bring in ideas for new products, the experienced people rarely saw the value in them. The prevailing belief was that those who were younger were there to learn and assist but not necessarily to innovate. After an idea would be rejected by older bosses up the management chain, a younger staff member would say, "Okay, I'll just go to Kickstarter."

This seemed to me a perfect example of the way established organizations can overlook the tremendous insight employees, especially younger ones, can bring to the creation process. To be sure, recent graduates can lack experience and discernment, but there has to be a better strategy than encouraging young people to wait until they've paid their dues to start thinking about implementing their ideas. The cost of losing those ideas can be high. For that reason, companies should be thinking about setting up their own internal venture funds. Or if they don't have that capability, they should be bringing in experienced venture capitalists who can help turn ideas into new products.

Just as many companies are failing to capitalize on the vast amount of knowledge their employees possess, so too are many of us overlooking our own skills and experiences. At the risk of stating the obvious, I can see no better first step in improving your knowledge-mining skills than becoming familiar with the knowledge and abilities you already possess. Even if you feel you have a firm grasp on areas where you have an edge, you may be surprised at what a formal "audit" turns up. Set aside some time to do a proper "experience, knowledge, and skills" audit—it can be a bullet-pointed list, an essay, or a folder of photographs of your work or areas of interest. What's most important is that you don't edit yourself: that mother from New Mexico I spoke to didn't see her sewing ability as a creative skill, but it certainly was.

Keep in mind that your accomplishments may not seem as glamorous or impressive as those of other people. But they might be. For that reason, you'd be wise to enlist the help of a

close friend, a family member, or others who really know you—
teachers, coaches, managers, editors. The point of this exercise is
to begin to see knowledge and experiences you've probably taken
for granted as skills that can be applied to new endeavors.

Organizations, too, should have some system of auditing the
skills and areas of interest of their employees. This could be an
extension of the "20 percent time" policies that many companies
have already adopted. Some companies call this "free time," but
giving your employees the opportunity to devote some time each
week to projects they feel are important is anything but. It's a
valuable investment in your organization's Creative Intelligence.
3M was a pioneer of this strategy, launching mandatory 15 percent free time in 1948. It continues to be one of America's most
consistently innovative companies, aiming to derive a significant
percentage of profits from new products created in the previous
five years, a goal it achieves again and again. Products such as
clear bandages, painter's tape that sticks to wall edges to protect
against paint, optical films that reflect light, and sandpaper that is
so sharp it acts like a cutting tool all came out of 3M's 15 percent
free time program.

Google has had a 20 percent free time policy since its founding, a program that had been extremely successful in generating
different kinds of services—at least until competition with Facebook led cofounder Larry Page to mandate that everyone use his
or her 20 percent time to work exclusively on one topic: catch-
ing Facebook by building out Google+, a social media site where
people can meet and exchange images and information.

Past successes at both Google and 3M suggest that the freedom to focus on whatever employees choose has helped motivate
them to come up with ideas they might not have had were all
projects simply assigned to them. For managers that are still con-
cerned about the viability of such programs, they can incentivize

20 percent time to encourage employees to connect their passions to projects that actually have a chance at creating value for their organizations. Assess the program's success six months or a year in: Has it led to any promising developments? If it hasn't (and I doubt that will be the case), all you've lost is a few Fridays that people might have otherwise spent phoning it in, making plans for the weekend, or watching cat videos on YouTube.

Another strategy some companies have adopted is mapping the informal "creativity networks" inside their organizations. Most scientists, engineers, and designers naturally talk and work with people across company boundaries. They often go outside their companies altogether to link up with people doing similar work in different silos. Accepting and embracing these unofficial networks of creativity, even if these teams cross bureaucratic boundaries, can be hugely productive. Companies should be nurturing their "bandito creatives," men and women who've already demonstrated a willingness to break the conventional rules if that's what it takes to do their jobs creatively. Find them and map their networks. Ask them questions: Who are you talking to inside and outside the company? Who did you work with on this research or project? Who are the most original thinkers that you know of? Where do you go for inspiration? And, most important of all, encourage this "off-the-books" engagement because it just might lead to that new $1 billion product.

Universities need to encourage this kind of creativity mapping as much as corporations. The best new research tends to come at the intersection of two or three established disciplines— think bioengineering, neuroscience, digital fabrication. Mapping these networks and funding the gutsy scientists who reach across disciplines is one way universities can keep at the forefront of research.

On a personal level, we can all implement our own 20 percent strategies, committing to a certain amount of time each week to pursuing areas that interest us, even if we can't see the clear value of those activities yet. We can spend some of that time "mining the past" of industries that interest us. We should be setting up at least one meeting a week with an expert in our fields and asking: What's currently on his or her mind? What's exciting and inspiring these days? What's worrying?

We can also ask experts what's *not* there. What do they think people in their industry are missing? What is everyone getting wrong? What should they be thinking about or focusing on instead?

Cultivating all of this knowledge is a good starting point, but if we want to use it for creative purposes, we need to actually start thinking about how to put it into practice. If you've picked up this book because you're hoping to solve a particularly vexing challenge at work, you would do well to close it right now and go for a walk. Actually, read the next two paragraphs, then go.

These days, people are constantly connected: texting, e-mailing, talking, and responding from the moment we get up in the morning. Try disconnecting for a bit of the day, especially in the morning. Take a walk. Walking alone is an excellent strategy for freeing your mind up so that you're better able to bring together different areas of knowledge.

You can keep a particular challenge in mind as you walk, or you can just look around and see what other inspirations strike you. Steve Jobs was a walker. Mark Zuckerberg is a walker. IDEO cofounder Bill Moggridge talked about walking the High Line in New York to find clarity and creative inspiration. Walking to a local park (or nearby beach, if you're lucky) or even just around your neighborhood can give you the space you need to start mining the knowledge you've accumulated and connecting dots. And finding that neighborhood coffee shop to hang out and just think is important too.

The late Pulitzer Prize–winning reporter and professor Donald Murray once wrote, "My writing day begins about eleven-thirty in the morning when I turn off my computer and go out to lunch. I have written and now I will allow the well to refill." Notice he said his day begins when he turns the computer *off*. We've come to equate time spent sitting down in front of our computers with working when we forget that our minds are still going, learning, making connections long after we've left the office. Unfortunately, it's easy to forget the important role that reflecting, processing, *not doing*, plays in creativity.

Okay, you can go for that walk now.

The final thing to remember when it comes to Knowledge Mining is that it's going to be very difficult for all your hard work learning, mining, and connecting to pay off if you don't spend time thinking about how to connect to what people find meaningful. It's important to translate your enthusiasm and insight into a narrative that others find compelling and want to engage with. In this economy—in any economy, really—this kind of engagement is key.

And so, in the next chapter, I'll introduce a competency that's all about rethinking the many ways we engage, reframing the stories we tell about the world around us, and possibly even reimagining the future.

4

Framing

CHARLES ADLER LOVED MUSIC SINCE he was a kid. But, at the encouragement of his father, he went to Purdue to pursue a mechanical engineering degree—and pretty much hated it. Luckily, it was the mid-nineties and this thing called the Internet was taking off, so there were a number of options for someone who didn't want to follow a traditional educational path. Adler fell in love with graphic design, dropped out of school, taught himself how to code, and began doing interaction design for media consultancies in Chicago.

Adler honed his skills designing great user experiences on the Web, becoming Director of Strategy and Information architecture at Agency.com, a media market consultancy, and cofounding SourceID, an interactive design studio. He also cofounded *Subsystence* magazine, an online art publication that offered free music downloads. His interest in that project was, perhaps, a giveaway—he really wanted to be part of the music and art worlds. After a number of years, he was ready to quit.

Perry Chen was ready to quit too. In 2002, the New Orleans–based music promoter had grown tired of going to the same handful of people time and again to raise money for concerts. It had become too difficult and too risky. Chen was thinking that perhaps he could put on concerts by harnessing what was then a new phenomenon, crowdfunding. If he could get a lot of people to put up small amounts of money, they would act not only as audience members but also as funders.

There was a model for the idea: A Chicago-based community of T-shirt makers and buyers called Threadless was experimenting with a new kind of retail—designers put designs online and people voted for their favorites. Clearly, the line between the traditional roles of producer and consumer were becoming blurred.

Chen was living in New York when he met Yancey Strickler in a Brooklyn restaurant and laid out his idea for a crowdfunded music promotion site. Strickler, who was editor in chief of eMusic, an online music store, at the time, liked the idea, but neither had any expertise in actually making a website. So they began asking around. Chen's college roommate recommended Adler, who, by then, was ready to make the leap into something new. When the three met in New York, they asked themselves: How do we turn consumers of music and art into patrons?

It was a very simple question, but the answer changed the story of how we see and engage with music and art. In so doing, it may have also changed the way all start-ups are funded, ushering in a new culture of entrepreneurial capitalism. They called their new venture Kickstarter.

Adler designed an easy-to-use, engaging site for Kickstarter where people present their creative ideas to a small panel that curates the offerings and passes their selections directly to an online audience, termed "patrons." Then the patrons vote by either doing nothing or pledging money.

If people don't pledge enough money to reach the minimal goal set by the artist, they get a refund. If the amount pledged exceeds the request, the artist or musician gets all the money as well as 100 percent of the ownership. All transactions are handled through Amazon.

On one level, Chen, Strickler, and Adler built a crowdfunding model. On another, they developed a *crowd-building* site, as Kickstarter not only raises funding but also creates an audience— and a uniquely invested audience, at that. When you pledge money to a project that's launched, the creators take you along for the ride, e-mailing and chatting with you, bringing you inside the process of creating. Which materials should we choose? Where should the product be manufactured? How should it look?

At the end of the project, the creators have a committed audience for the product, and audience members spread the word to their friends and families. There is a tiered level of patronage, from a few dollars to thousands. Everyone gets something, from a token "thank you" to the actual product to the product plus dinner with the creators. Investors, or "patrons," help create something new and perhaps important. The experience is not a simple transaction like buying a work of art at a gallery or an album off iTunes. It's deeper and richer.

The first Kickstarter projects raised $1,000, $5,000, even $25,000. But the audience for participating in the creative process proved both larger and more willing to invest than the founders had imagined. Between its launch in 2009 and October 2012, successful Kickstarter projects raised a total of $316 million, mostly for art and music. If Kickstarter continues to grow at this rate, it will soon rival the National Endowment for the Arts, which had an operating budget of $146 million for 2012. But Kickstarter doesn't finance just art and music. A campaign for new watches based on the iPod nano music player (you snap it into a special wrist band) raised nearly $1 million; the resulting products, Tik-Tok and LunaTik, sold tens of thousands for their designer, Scott Wilson, at his company Minimal in Chicago. Printrbot, an inexpensive 3D printer, raised $830,827. A San Francisco–based studio raised more than $3 million to create Double Fine Adventure, an online game.

Kickstarter changed what it means to be a creator and a capitalist, a maker and a patron. Chen, Strickler, and Adler built a site, and a community, that makes us all potential participants in the process of creation and gives us all the potential to become entrepreneurs. Kickstarter broke down the hierarchy of the traditional investment model and the elitism of being a creator. The same idea is at the heart of the JOBS Act, new legislation that permits people to invest directly in start-ups and receive equity participation. It's a model that promises to democratize investment, making all of us venture capitalists.

By changing a narrative that we all accepted as true, perhaps inevitable—there are creators, there are financers, and then there's the rest of us—Kickstarter is one of the great "reframing" stories of our time.

WE ALL HOLD A NUMBER of beliefs about the world that color our interpretations of people and events. These "frames" are there, whether we are aware of them or not, and people with Creative Intelligence have a knack for turning frames a bit to the side, spinning them around, or maybe even tossing them all together, changing the way we view the world and our place in it.

Gregory Bateson, a British anthropologist and Margaret Mead's third husband, is known for originating and applying the term "frame" to the field of sociology in his book *Steps to an Ecology of Mind*, in 1972. In it, he talks about how he developed his ideas by observing monkeys in a zoo in San Francisco. When monkeys interact, as they do all day long, their playful behavior looks very similar to the way they act during a fight. In order to survive, monkeys use a set of expectations about the behavior to interpret what's happening. These expectations, as well as the behaviors of the provoking monkey that signal "play" or "fight," are examples of "frames." Frames are more than mere stories. They provide a schema of expectations that can help us interpret the meaning of a situation, the intentions of the players, the consequences of the outcome. A frame gives us the answer to the question "What's going on here?"

There's perhaps no better story to illustrate framing than that of one of Pablo Picasso's most iconic "paintings." Picasso spent his career moving in and out of different styles of painting, constantly absorbing information and techniques, frequently working with other artists, always moving ahead to create something new. He was influenced by Gauguin, Cézanne, and African tribal mask art. He painted side by side with Braque in 1908 to construct an entirely new art movement, Cubism.

But then he went further. In 1912, he crafted a deconstructed Spanish guitar out of cardboard so that it resembled a Cubist sculpture. He called it *Still Life with Guitar*. When Picasso brought art critics to his studio to see his innovation, they were appalled. What was it? A painting? A sculpture? They refused to "frame" it as painting or even "art," and he never showed it to the public in a gallery or a museum. Yet he clearly loved it, as he went on to make a second piece out of ferrous sheet metal in 1914. In so doing, Picasso successfully changed our expectations of what belongs inside a frame. In 2011, the Museum of Modern Art in New York exhibited both of Picasso's guitars by putting them up on walls. Clearly, Picasso had reframed painting.

Ninety-four years later, a young artist from Laguna Pueblo in New Mexico began to experiment with her own reframing of traditional paintings. Marla Allison won the first Innovation Award at the Santa Fe Indian Market for *Mother*, a portrait of her mother sitting in a chair, legs crossed, wearing sneakers. In the top right quadrant of the picture was an LCD screen that played an interview with her mother, who described her life growing up as a Native, and moving, as a child, into the detention camps that housed Japanese-Americans during World War II. Using technology, Allison enabled the object of the portrait to actively participate in her story. Allison broke the conventional frame of the portrait.

Erving Goffman, a Canadian-born sociologist, moved Bateson's work forward by arguing that people actively construct their frames. Like actors on a stage, he posited, we all have the ability to create new roles, new story lines, new engagements. His observations are particularly useful at a time when industries that once felt safe are crumbling, and the jobs we worked so hard to master are disappearing. Teachers, doctors, soldiers, advertisers, factory workers, book editors, salespeople, designers, CEOs, nonprofit managers, even politicians all realize that we need to change ourselves and our institutions. But how? It's an enormous effort, and it's terrifying. How do we challenge the conventional ways of doing things, even if we know they aren't working? How do we break out of our own ingrained habits and dare to try something different?

We can begin by understanding that we interpret the world through specific lenses—and it's within our power to change our glasses. During the era of money culture, when Wall Street was riding high, nothing was cooler than working for Goldman Sachs. But today, thanks to outrage at the 1 percent by members of the Occupy Wall Street movement and the Tea Party alike, "the new status jobs aren't at Goldman Sachs," said a former Goldman analyst who left Wall Street. "They're at Google, Apple, and Facebook."

By practicing framing and reframing, we can begin to see ourselves behaving in alternative ways. We can reframe what our industries and institutions actually do and how they do it. Designing new industries and careers that go with them is not simply a matter of perceiving them differently. But having the creative confidence to challenge established modes of organization and behavior is a necessary first step toward change.

NARRATIVE FRAMING

Chemotherapy is a difficult therapy and an uncomfortable experience at best. The story is well known. If you are unlucky enough to be diagnosed with cancer, your treatment will likely involve spending a lot of time in a large room alongside a bunch of strangers with IVs in their arms dripping chemicals into their bodies. Soon you feel nauseous and weak; that's when you have to get up, leave the hospital, and go home, getting closer and closer to vomiting along the way. That's the way it's done. That's the way it has always been done. That's the story of cancer treatment—or at least it was until someone reframed the story.

Like a number of industries, health care is hierarchical and centralized, but safety concerns make for even more rules and regulations than most. When you're sick, you go to a big hospital, wait there, get treatment, and go home. For many situations that require specialization and expensive equipment—delicate operations, advanced treatments, intricate diagnosis—that approach makes sense.

But for other illnesses and treatments, the existing framework of health care is ill-suited, unnecessary, and expensive. For regular chemotherapy appointments, that's especially true. In 2010, Irish Maliq, the Project Manager of Strategic Planning and Innovation at the Memorial Sloan-Kettering Cancer Center, decided to do something about it: Maliq gathered students from NYU, Carnegie Mellon University, IIT Institute of Design, and Parsons The New School for Design to reframe the chemo experience.

The first thing the team did was observe the patients getting chemo treatments and ask them how they felt about it. It became obvious that people took cars or subways long distances to get to the hospital and then had to return home just as they began to feel sick. Many felt the panic of being trapped in a car or on public transportation while on the verge of throwing up. Others just hated the sterility and lack of privacy in hospitals. Almost all of them wanted treatment closer to home.

So the students reframed the chemo experience, designing a space with warm colors, Internet access, and coffee. It was small, with just a few rooms, and located in a Brooklyn neighborhood close to where many patients lived. The team designed new lounge-like chairs for treatment. Nurses didn't wear formal uniforms, just regular clothes.

In the end, the neighborhood chemo centers proved to be less expensive than bringing people to the hospital, and patients liked them a great deal more. Now Sloan-Kettering is doing research on outcomes—whether or not the changes actually improve recovery rates.

Reframing is about breaking routine. We tend not to think about the narratives that structure our lives. We take the frames we were born into—and even many of the ones we've created ourselves—for granted. In a world that doesn't change much, in a time of relative stability, the need for reframing is not quite so urgent. But if you are enmeshed in narratives that don't work, changing your frame just may be the first step toward changing your life.

Who's Telling the Story?

One of my first cover stories for *BusinessWeek* was called "I Can't Make the !@#&%! Thing Work." It was 1991, and I was frustrated about how complicated my VCR was, and feeling guilty (as a consumer and a man) at my inability to operate the latest, coolest technology. I really wanted to be able to time-shift my TV programs, but it was such a complex process, with so many buttons to push, that I failed practically every time. It was infuriating.

I did some research and found that most consumers of VCRs at that time felt the same way. The product was too complex for the average user. And the manuals were practically unreadable. Yes, they had been translated from Japanese to English. But I discovered an even bigger translation problem: The manuals were written in the language of engineers, not the typical consumer.

The story of the first VCR technology coming from Japan is one of brilliant Japanese designers designing brand new gizmos for their buddies, then selling them around the world. Engineers naturally placed importance on advanced technology and complex functionality. Early adopters, engineer-types, and geeky teenagers (at the time, that meant mostly boys) understood the story, but the rest of us were left fuming. I'm sure other engineers could see the logic of the machine's many functions, but the description wasn't helpful to me and millions of other consumers. The engineers at Panasonic and other early VCR makers couldn't get out of their own way to see how their products were actually being used by average people.

One Japanese company was able, for a time, to do just that. Sony began making VCRs with only a few buttons, instead of a hundred. Buttons for the most important functions—the Power, Play, and Rewind buttons—were larger and more prominent than on most VCRs. All the other more complex functions were there too, but you could simply avoid them. The engineers and designers—led by then-CEO of Sony, Akio Morita—were able to envision a new narrative frame for the VCR. Instead of being a technological marvel, the Sony VCR would be a great consumer experience.

It's no accident that when he launched his company, Steve Jobs followed the Sony model of a consumer-friendly technology company. He knew that it was always tempting for technology companies to frame themselves from the engineer's point of view, not the consumer's. Over time, Jobs would fight his engineers constantly to keep Apple products easy to use.

No one has reframed the story of personal technology quite like Steve Jobs. With every new product, he further moved the focus away from engineered functionality and toward user experience. His contributions transformed not only the personal computer market, but the entire field of design. Jobs, who framed his very role in his own terms, thinking of himself as a designer, not an engineer, was able to see a completely original relationship between people and technology. Apple under Jobs transformed entire industries and the way we interact with products on a personal level.

Frames affect not only our interactions with products, but our beliefs about the world. In his 2004 book *Don't Think of an Elephant!* Berkeley professor of cognitive linguistics George Lakoff argues that much of the differences between the two major American parties can be attributed to their different frames, and the strength of their convictions about them. Democrats see government as a "nurturing parent," a kind, helping organization that enables people to do better. They see the world as a relatively safe place that people can thrive in if only given a little help. Within this frame, paying taxes is a social obligation that we share.

Republicans, however, see government as a stern and disciplinarian father that needs to toughen up weak individuals so they can be strong enough to survive in a hostile world. In the Republican frame of government, people must rely on their own resources. Taxes are a burden and regulations interfere with independence and freedom and need to be cut back to the minimum.

Lack of awareness about the frames that color our perceptions of the world severely limits our ability to see new opportunities. Yet one of the first steps in creating something new is to break free of the old definitions in order to interpret facts and patterns in new ways. And that can be quite difficult.

For companies, this can often mean a shift away from an area of focus that defined the company for decades. IBM, for example, found itself in serious difficulty in the 1980s because of a decreasing need for its core product—mainframe computers that processed lots of data. But under a new CEO, Lou Gerstner, who came from outside the computer business as CEO of RJR Nabisco, IBM was able to reframe itself as a consulting and service business that could help companies solve problems using data and analytics. This reframing helped the organization prosper over the next two decades.

Other companies have been less successful at reframing themselves in turbulent times. Kodak was slow to rethink the definition of photography in a digital age. It lagged behind its consumer base—and a slew of competitors—that moved faster than it could or would. RIM, the maker of the BlackBerry, once the premiere smartphone on the market, was not quick enough to shift its strategy from a focus on corporate clients to average consumers, a demographic that Apple was able to swoop in and capture. Google today finds itself in the position of so many companies before it: It's long been dominant in one area—search—but it's investing a ton of resources into coming up with a way to rival Facebook and other social networking sites. Whether it will succeed, or change course and reframe itself in another way, remains to be seen.

Reframing can open up enormous opportunities. You begin to see yourself in a different job or career. You begin to see new markets where before there were none. C. K. Prahalad, Professor of Corporate Strategy at the Ross School of Business in the University of Michigan, reframed the concept of "poverty" by not viewing it as a "condition" that was very difficult for people to exit, but seeing the poor as consumers with money, just not much of it. Global corporations like Unilever responded to this line of thinking by making small sachets of shampoo and other products sold in ones and twos to people who could afford them for the first time. A wave of innovation followed, from cell-phone-based banking to low-cost eye care.

We talk a lot about the power of story these days—storytelling techniques aren't just embraced by artists and performers, they're used in a wide variety of industries to do everything from teaching unfamiliar concepts to motivating staff. But stories themselves exist within a larger framework—their meanings are dependent on the way they're being told, who's telling them, and why. Once we understand that, we can begin to reframe not simply our stories, but also the way we present ourselves and our organizations. The most important reframing of my career came when I transitioned from *BusinessWeek*'s editorial page editor to head of the Innovation & Design online channel and blogger. My new role was just one example of the massive reframing of media and journalism. No longer was it enough for those of us in the media to simply tell people what to think about important issues (and what the important issues were); there was a widespread desire to participate in conversations—sometimes leading, sometimes following—and in my new role, I learned a great deal from commentators and contributors who, in previous times, would have simply been readers.

Is a journalist just a person who works for an established publication? Is a college professor just a teacher who delivers insight from a podium in a classroom? Is a capitalist just a manager in a big corporation? Asking ourselves these kinds of questions can help us rewrite our own narratives—and perhaps the narratives of others around us.

ENGAGEMENT FRAMING

A decade or two ago, someone in the media might have a couple interactions in one day: an interview or two, a talk with an editor, a morning meeting with a team.

Compare that with today when that same person might peruse hundreds of Tweets, use Facebook to interview five people for a story, Skype with a team in Asia, write six short posts for his or her work blog and personal Tumblr, and still have that morning news meeting.

Thanks to social media technologies and globalization, the typical American, Brazilian, and German now actively engages with a larger number and a much more diverse group of people than ever before. We used to define our identities by the groups and communities we belonged to. Now we actively create our own communities and define our interactions within them. When we have tens, hundreds, even thousands of interactions each day, it makes sense to constantly "check in" to how we're framing ourselves, our behavior, and our conversations.

Frames have always affected our conversations. You behave differently with your parents at a party than you do when chatting with friends online. Companies have different ways of communicating with their suppliers than they do with their customers. But now that we find ourselves bombarded with information from every angle, examining the many engagements in our lives can lead to the creation of deeper, more meaningful interactions. What does it mean to have five thousand friends on Facebook?

What is the true cost of "free" music, art, and entertainment, of promotional "gifts"? My older, senior students are finding the interactions with most of their "friends" too shallow, and are culling their Facebook networks down to a couple of dozen "real" friends. They love the idea of information being "free" on the Internet, but expect it to come with hidden obligations, like unwanted ads. As for "gifts," they like Groupon coupons that give them half price for restaurants and stores, but feel no obligation to return the gift by going back again.

Businesspeople have long embraced the idea of "community," but that doesn't mean they understand what's meaningful about belonging to networks like Kickstarter or movements like the Tea Party or Occupy Wall Street. We now share more deeply, more widely, and more publicly than ever before. And we have many more roles to play. Understanding how people frame their engagements allows individuals and organizations alike to be more creative on their feet, more deftly navigating the sea of interactions we have each day, both traditional and modern.

Consider the many kinds of engagements one person can have through Kickstarter. In 2011, I pledged $25 for a proposed photo book of the Lower East Side of Manhattan. I thought it would make a great present for my mother, who grew up there. So I began as an investor, but as soon as the author started posting updates each week to all of us who had pledged, I also became part of a new community. Those who offered advice on the book became consultants. When the book came out, I got one, becoming a traditional consumer. In this one "transaction" with Kickstarter, I had the opportunity to engage in half a dozen different ways.

The ability to rethink engagements is a hugely powerful technique for creating new ideas. Take restaurant reservations. Before OpenTable and other apps like it, we had to call each restaurant sequentially, wait for an answer, deal with often snooty gatekeepers, and negotiate a specific time to dine. It was a time-consuming and sometimes anxiety-inducing experience. By taking the process online, Open Table allows patrons to check out many restaurants faster, see what time slots are available, and book one without having to beg.

Wherever there is a point of interaction, there is potential for innovation. But the "right" engagement will vary from group to group, industry to industry. When we're having problems communicating with people from different cultures, our default "solution" is to speak louder or repeat ourselves. Corporations echo this behavior on a grander scale with loud, expensive marketing campaigns or one-size-fits-all customer service strategies that fail to take into consideration the ways people really want to engage. Language barriers aside, the better solution is to investigate how a particular group or culture prefers to engage.

You Need to Knock

Engagement frames vary from generation to generation. I was on a jury in Toronto judging student projects where this became clear to me. One of the wining projects, KNOCK, was based on the insight that most people aged eighteen to twenty-six don't like to make direct phone calls. Calls are considered an invasion of privacy and too intimate for a generation that would much rather interact by texting. So you need to "KNOCK."

"KNOCK" was an app that allowed you to request a real-time call via text message. You KNOCK on an acquaintance's cell phone "door," say who you are and why you are calling, and offer a choice of times. You can also KNOCK the message forward and connect to others. The students said KNOCK would help struggling young professionals who are unfamiliar with phone etiquette, as well as people generally lacking in social graces.

Each group, whether it's a demographic or a regional or a national group, has its own culture—the groups we create are no exception. Each has its own rituals, shared values, community expectations, and secret "knocks." Learning what they are is a way of saying, "I respect your values." It's also a promise that you have something that might be of value to others.

Food is often the way to "knock" and engage a community. We all bring wine or a dish to thank friends or family for hosting. We eat with new friends to ritualize and legitimize our engagement, even if we don't really like the particular dishes. I was in the Peace Corps in the Philippines in 1968 when I went up into the Cordilleras mountains in Luzon to see the rice terraces of Banaue. Riding on a motorcycle, I wandered into a village of the Igorot, a highland people. A local chief asked me to stay to eat. Though I was eager to see more of the rice terraces and get back to Manila, this was the land of the Igorot and I was a visitor, so I accepted.

For lunch, they brought out a live monkey. The men, women, and children around the big table laughed at my wide eyes and open mouth but insisted it would be fine. One of the men took a small, sharp ax and removed the top of the monkey's skull. Another took a spoon, scooped out the monkey brains, and offered it to the village guest, me. This was the "knock" to their community, and I could have refused. The monkey was screaming, the brains looked slimy, and all eyes were intently on me, so I was feeling a bit afraid (the Igorot had been fearsome warriors and headhunters just two generations back). But I understood that eating the brains was my port of entry into Igorot acceptance, and so I did. The village exploded in laughter, embraced me as an honored guest, and sent a guide to show me parts of the mountains rarely seen by outsiders.

You can learn the rituals of engagement for any community and perhaps master the most important of them. In doing this, you are, in a sense, using a key to unlock a closed door and gain entry for yourself.

Lenovo, the Chinese computer company, had its own experience learning how to knock when reaching out to rural Chinese customers. The rural PC market in China is growing fast. Incomes of farmers are rising and so are the incomes of the young people who leave to work in factories that make products for exports. HP and Dell have sought to engage its rural Chinese consumers by competing on price. This is classic Western marketing strategy. What could be better for consumers than getting exceptional value at a low cost?

Lenovo, however, used its knowledge about Chinese rural culture to "knock" on their doors with another, more meaningful message. According to *Fast Company*, in rural China, people frequently buy PCs as wedding gifts. A computer purchase is not merely a market transaction, based on price; it's a gift that provides the foundation for a lifetime of social interactions between two families. So the box, the packaging, the entire presentation, is crucial. By making the box representative of the gift it contained, Lenovo was able to capture more of the rural market than HP—a knock worth hundreds of millions of dollars in profits.

Perhaps the most surprising gift you can give to customers is a creative approach to fixing a problem. Mistakes are actually perfect opportunities for reframing engagements. This is especially true in service industries that rely less on physical products than on interactions. That value can simply be an effort to correct a mistake and a commitment to saying, "I'm sorry."

At a Design Management Conference in 2010, I heard Kathleen Taylor, CEO of Four Seasons Hotels and Resorts, talk about the opportunities presented through glitches: "You know you are going to have mistakes in our business. The product is delivered through millions of interactions. For sure, something is going to happen."

Spills are an example of one of those inevitable mistakes. But Four Seasons managers and employees reframe the glitch, focusing not on the spill but rather the response. If there is a spill, for example, staff members try to dry-clean and return a jacket or coat while people are still eating. They apologize and move fast to remedy the situation.

The Four Seasons approach reminds us that the horror isn't in the actual mistake; it's in what happens next. If you reframe a mistake as a positive experience, people will remember it. If you don't, they'll remember the accident.

Give People Opportunities to Act

Throughout society, we are seeing our frames of engagement shift from passive to active, transactional to relational, impersonal to very personal. DonorsChoose, for example, is turning philanthropy into an intimate engagement by directly connecting individual donors to teachers who need school supplies. Kiva is directly connecting microlenders to microborrowers. Kopernik links new technologies for poor countries with corporations and individuals willing to offer capital. And Sparked allows busy people to "microvolunteer" by choosing from a number of causes, then completing various tasks—grant-writing, filling out surveys—from their own computers in ten-minute intervals. These new interactions and connections provide people with a deep sense of meaning. We can now do good and share our engagements with a larger community of our choosing.

Given the chance, we can build better engagements. Several years ago, I asked my students where they went for their creative inspiration. Out of a class of eighty-five, none of them said "mu-seums." Shocked, I sent them out in teams to research how they might build better museum experiences for themselves (and their generation). One group, led by Chia Schmitz, then a Design and Management senior at Parsons, concluded that going to a museum today broke three of their generation's cardinal rules. First, the prerecorded headphones containing information about the exhibits broke the sociability of the experience. Headphones were too isolating for young people who went to galleries and movies as a group and expected to share the immediate experience. Second, the information you could get on the exhibit was very limited and static. You could only read the small boxes on the wall next to the paintings or exhibits and listen to the tape recordings. "We're a search generation," said Schmitz. "We may not read books but we love information. And we love to keep it and share it." Schmitz's team also observed that nearly two thirds of the visitors to the two top museums they visited were not Americans. Nonnative speakers naturally had less-than-optimal engagements with the museum exhibits. What to do?

Schmitz's team proposed building an app before an exhibit even opens that provides a rich network of links to visual and print information on the artist or artists involved and all their work. It would be made available in most of the languages of the people who most frequent the Met and MoMA—French, Italian, Spanish, Portuguese, Mandarin, Russian, German, and, of course, English.

The team suggested that the smartphone app replace the tape recordings that plugged into your ear. The smartphone app wouldn't isolate you the way headphones did, and also would allow you to diary—to take photos, write down impressions, connect to other artists and sources. You could then share your museum experience with other people and keep it forever. This engagement with a museum, this experience, would be more lasting, informative, and meaningful than simply buying a book about the exhibit.

It would also be more participative. More and more people are rejecting the passive consumption of the past, whether watching TV shows at their official times, standing in front of a painting for a minute as a line moves past it, or accepting the diagnosis of a godlike doctor. We want to be actively engaged, and we want to shape that engagement. In conferences, we Tweet as experts speak at the podium, asking them questions, challenging their assertions. At a Design Indaba Conference in Cape Town in 2010, I saw Martha Stewart lose her audience in a blizzard of Twitter criticism about her delivering a speech that failed to address South African design issues.

Entertainers and politicians alike use Facebook and social media to communicate with their audiences and build brand loyalty. Businesses are trying to engage directly with their customers without going through media and market advertising intermediaries. Many brands have excellent stories that you want to be a part of (the "safe" ride of Volvo, the "thrilling" ride of BMW, the "rough" ride of Jeep). But an engagement frame that ties an audience directly and continuously to an organization may be even more powerful. It's a bit like the difference between watching a family on TV and being a part of an actual family. eBay is a family of strong engagers. The *Economist* is turning its readers into an active family by creating an Opinion Leaders' Panel through its Economic Intelligence Unit. The panel had more than 120,000 members in 2012.

Our drive to increase our participation and engagement with everything around us extends increasingly to our stuff. Smartphone users can scan QR codes, now pervasive in magazines, billboards, labels, and even the things we buy, to directly connect to a website or some other online source of additional information.

Just wave your smartphone in front of the QR code to connect. Innovators from a number of different fields are embracing the codes.

Pat Pruitt, a metalsmith from New Mexico, won the Innovation Award at the Santa Fe Indian Market in 2011 for his concho belt that used QR code as decoration. Each of the nine conchos had its own code, stylistically inscribed as an abstract piece of art. When you scanned it with your phone, a line of a poem written by Pruitt appeared, as well as a link to Pruitt's website.

There are billboards for Calvin Klein with no picture, no model, no advertising—just a huge QR code and the words: "Get It Uncensored." When you scan the code, you obtain access to a racy video of models in Calvin Kleins. Like a secret handshake, QR codes are an exclusive form of engaging.

The art world, too, is embracing the use of QR codes. In 2011 MoMA featured an exhibit called *Talk to Me*, curated by Paola Antonelli, Senior Curator of Architecture and Design at the museum. We're used to museum exhibits that feature beautiful and sometimes provocative things hanging on walls or projected onto screens; our engagement is typically limited to our response to what's in front of us. But Antonelli's exhibits focus more on our interaction with the objects around us. Nearly all the items in *Talk to Me* had QR codes that gave people a chance to interact more deeply with the exhibit.

In their 1999 book *The Experience Economy*, Joseph Pine and James Gilmore argued that we were evolving from an economy that valued things to an economy that valued experiences. While that was an important step toward focusing on what was meaningful to people, the term "experience" implies a passiveness that won't fly with a population that wants to create its own engagements. Engagement Framing is a more dynamic, active way of understanding how we interact, and how we might open doors that otherwise would have remained locked. It's time we recognized that today what we often value most are these special, active engagements.

WHAT-IF FRAMING

Asking "What if?" is the final, and perhaps the most exciting, kind of framing, as it's a way of challenging our understanding of the world as it is, and can be. In the process of thinking about a different future, you can increase your understanding about the present as well. Sometimes What-If Framing leads to radical new products or industries—sometimes what's important is the learning process. By pushing yourself out of your conventional narrative and engagement frames to think about what might be, you are forced to look deeply at your habits, biases, and beliefs.

What-If Framing is radical blue-sky framing, a provocation to go beyond what's known. Your aim should be to challenge yourself to think about entirely new possible options and how to create them. Asking "What if?" pushes you to think about the improbable, if not the impossible. Scenario Planning, War Gaming, and Futures Research are all powerful methodologies that involve looking at current trends to predict possible outcomes, and prepare for them.

A good strategy for What-If Framing is to begin by looking at major trends. GBN, the scenario-planning arm of the business consultancy Monitor, designed and facilitated a workshop on the aging population of China, Japan, and Korea at the Stanford University Center on Longevity and the Walter H. Shorenstein Asia-Pacific Research Center in 2009. The goal was to anticipate changes in culture and policy that might appear as a result of the sharp fall in fertility and the extraordinary increase in longevity.

According to GBN's Bulletin, China's sixty-plus population is soaring; by 2050, there will be 334 million people sixty-five or older in China—that's about the size of the US population today and will constitute about a quarter of China's total population in 2050. Japan already has the fastest-aging population in the world, but by 2050, one in six Japanese will be eighty or more years old. And South Korea, at 1.2 children per woman, has one of the lowest fertility rates in the world. By 2050, 35 percent of its population will be sixty-five years old or more.

Nearly all the experts on population aging and policy who were interviewed for a report given at the conference expressed concern about how to care for the aging population. The scenarios and predictions ranged from a return to "traditional values" to increasing Westernization. One interviewee envisioned a future in which machines played a greater role in care and service for the elderly; another reflected that because the aging population desires more independence, a more Western approach to care could be likely to follow. And more than one expert argued that because government alone cannot provide all the necessary care for this growing population, there were opportunities for the private sector to set up creative, local services.

This kind of looking forward will be familiar to many who've had corporate strategy experience, but What-If Framing needn't be as formalized as scenario planning or war gaming. Lisa K. Solomon is a principal at LKS Partners, a consultancy that specializes in leadership training and scenario planning. She is also a professor at the California College of the Arts MBA in Design Strategy Program. At a Design Management Institute conference in 2010, Solomon said, "There is usually paralysis in the face of uncertainty. There's a denial of the future."

Scenario planning is one way to push past that uncertainty. These scenarios, however, have to be structured and focused, according to Solomon. First, she recommends always taking the long view, at least ten years out—twenty to thirty, if possible. All of us, from the working parent to the corporate financial officer, are driven by near-term horizons. Our long-term goals can easily be eclipsed by day-to-day meetings, deadlines, assignments, or reminders of those quarterly profit targets. Three months appears to be a "long time," but it isn't. Planning our careers and plotting the course of businesses and nonprofits requires a much longer time horizon.

My students, for example, will often work for four or five different companies within the first three years after graduating. To most boomer parents and grandparents, that may appear to be frivolous and a waste of time. But they would be wrong. Look out ten or twenty years and you can see the wisdom of hedging your career bets by trying out different jobs and industries, taking advantage of the different opportunities that present themselves.

A second principle from Solomon is to think from the "outside in." Most of us spend our days doing what we already know. We keep our noses to the grindstone and focus on the matters at hand. We neglect the big structural changes that indirectly influence us over time. But that's a mistake if you're thinking about who you want to be and what you want to do in the years ahead. In the movie *The Graduate*, Dustin Hoffman's character Ben is taken aside by Mr. McGuire, who tells him to think about a career in "plastics," because it's the future. Ben, being young, simply stares at him, not understanding (or, perhaps, not wanting to). But it wasn't bad advice. Anticipating the big technological, social, demographic, economic, and political changes can provide important insights as to where you might channel your creative energy (or existential angst). The aging of the boomer population has led to many people entering careers in training, coaching, and medicine to better serve their needs. Given the desire of grandparents to keep in touch with their families, I'd bet that providing technologies or services that facilitate those connections will be a thriving sector in a decade or two.

A third way to strategize about the future, according to Solomon, is to "encourage multiple perspectives," opening ourselves up to collaborations with people outside our peer group or culture. Consultancies such as Ziba, IDEO, and Smart Design bring together designers, anthropologists, psychologists, historians, and linguists from different cultures to form teams that see things from different angles. Working with people whose views and experiences differ from your own can challenge you to look wider, toward the periphery, at many possible outcomes.

Venturing Past the Possible

While there is a practical element to What-If Framing, at its best the strategy challenges us to imagine the unimaginable. My favorite sources of this kind of What-If Framing are science fiction writers. They can shape stories about the future that take our current capabilities and expand upon them or simply invent scenarios that become real possibilities. Arthur C. Clarke was the quintessential science fiction writer in that regard. Long before *2001: A Space Odyssey*, the science fiction classic he created with the director Stanley Kubrick in 1968, he made what is probably his most important prediction. According to Neal McLain, in a 1945 letter to the editor of *Wireless World* magazine, he propagated the concept of a satellite that would not spin around the earth but remain stationary in orbit as the earth turned. In a subsequent paper he wrote for the magazine, published in October of that year, he said that if "space stations" could be shot into space on rockets into a circular orbit 22,240 miles above the equator, they would match the earth's rotation of twenty-four hours. In that kind of orbit, a satellite would stay above the same spot on the ground and enable signals to be transmitted to it and back to earth. This geostationary satellite concept would, nineteen years later, become the basis of all communication satellites. It took that long for the United States to build on the German V2 to make a rocket, the *Delta*, large enough to launch such a satellite.

Clarke may well have written the mission statement for What-If Framing with the second of his "Three Laws" of prediction: "The only way of discovering the limits of the possible is to venture a little way past them into the impossible." Unfortunately, his ability to think about the impossible didn't extend to everyone else. In a piece entitled "A Short Pre-History of Comsats, Or: How I Lost a Billion Dollars in My Spare Time," Clarke said that a lawyer had persuaded him not to apply for a patent for his idea of a geosynchronous satellite. The lawyer, Clarke said, thought the idea of relaying signals from space was nonsense.

FRAMING: CHANGING THE WAY WE SEE THE WORLD, CONNECT WITH PEOPLE, AND THINK ABOUT THE FUTURE

If your life has been directly affected by massive upheavals to the economy or your industry, it's easy to feel overwhelmed or helpless. Even if you've managed to survive these shifts unscathed, the sheer speed at which change occurs can make it feel as if you're drifting along through life rather than steering it.

If you're just starting out, you may feel you have little control over your career path and your future—that the choices you made when you were nineteen have determined the way you'll be spending the next six to eight decades of your life. If you're more established in your career, your default behavior during moments of instability might be to keep your head down and work harder, hoping you can just make it to that next promotion, or to retirement, without your job or industry being affected.

While you may not be able to eliminate the uncertainty or turn back the clock to some time when jobs were "safe," basic framing skills can give you a way to reset your perspective of who's in control. Knowing that you have the power to frame your interactions and your place in the world opens up opportunities to interact with people in new and creative ways and rethink the way you see everything.

When you are encountering any change or challenge, it makes sense to take a moment to remember that you, and everyone around you, are always participating in a narrative—a story that contains some elements that you can change. Reframing is often as simple as asking, What's really going on? And how did we get here? It sounds simple, but how often do we really ask these kinds of questions? As our lives have become more virtual and participative, there's naturally less space for reflection, less time to think about what really matters.

Creative individuals understand the value of stepping back every once in a while and letting go of "the story." The story can be one you tell yourself about your abilities or, at the organizational level, a truism that prevents people in your industry from trying something new. What successful tech entrepreneurs seem to understand instinctively is the value of jumping on the part of an idea that works and abandoning what doesn't, but it's a lot harder for established corporations to reframe themselves if new technology and changing cultural values render their competitive advantages obsolete. Here's where framing comes into play.

One company that was able to successfully reframe its story during a time of hardship was Canadian transportation company Bombardier. Company founder Joseph-Armand Bombardier invented the first snowmobile to transport people across the snows of Quebec. But in order to stay solvent during World War II, the company had to look beyond its origins and begin making military vehicles.

Bombardier framed itself not as a snowmobile company, but as a company that moved people, and over the decades, Bombardier moved into trains, and then regional passenger jet planes. Leaders needn't abandon their organizations' identities during technological and economic shifts, but they do need to rethink how narrowly or broadly they define themselves. If Kodak executives had thought about photography as imagery, rather than just chemistry, they may have weathered the storm that came in the form of nimbler digital competitors.

When it comes to our interactions, whether in person or on the Web, it's helpful to remember two basic Engagement Framing principles: (1) our communication styles are not fixed. We behave differently depending on the circumstances and the people we're interacting with, and (2) we must be aware of the frames that affect our behavior (our biases and mental shortcuts, our habits and fears) if we are to successfully connect with people.

When people don't "get" what we're saying, we can ask ourselves, How can I express things in *their* terms? When we don't get it, we can ask ourselves, What am I missing? That means listening, *really listening*, not simply to what others are saying, but to what's behind their words. Why is this important? What's the backstory? Why is what they're saying meaningful?

The more knowledge you have about the world, the better able you are to frame. People who are seen as "wise" or "experienced" are often just those who are better able to connect with people whose frames differ from their own. But all the knowledge in the world cannot replace the importance of continually checking in to make sure you're on the same page.

Improving your Engagement Framing skills can help you have deeper, richer conversations—a key ingredient of any creative collaboration. But, beyond that, we can begin to reinvent industries by thinking about the ways that people prefer to engage. The rise in social media, for all its benefits, has had unintended consequences—burnout, people seeking more "authentic" experiences and real engagements. A longing for "social graces" or deeper, more meaningful engagement, is a common theme among my students. One group in class designed the most wonderful letter-writing kiosk, replete with beautiful papers and inks and even "starter" letters that showed people how to write them.

There are countless opportunities for people who can devise ways to help people slow down and organizations that can bring such ideas into being. Just as Sony came to dominate the market for consumer technologies by simplifying products, there is an increasing need for companies to step in and devise ways to help people simplify their lives. Ordering movies on demand on TV is a slow, clunky, frustrating process, providing an opportunity for Apple or another company to step in and create an easier, simpler experience. Had Philips, the Dutch corporate consumer giant, been able to reframe our engagement with television, it might have been able to save its flat panel TV business. The possibilities for reframing basic business and organizational models are greater than most of us think.

Take the story about the reframing of Umpqua Bank, one of Oregon's oldest banks, which has long served the local agriculture and ranching communities. By the early part of the twenty-first century, it was stagnating and had lost its way. Umpqua hired Ziba for help. The consultancy's first act was to ask the simple question "Who is Umpqua?" Ziba discovered the bank's roots as a general store and its history of serving rural depositors in the timber logging industry.

So Ziba redesigned Umpqua's branches as modern, friendly places to gather, have coffee, plug into the Internet, read the newspapers, and talk to some knowledgeable people about your banking needs. Even before you take your first step into the bank, Umpqua welcomes you. Because Portland is a big dog town, there's a dog dish full of water out in front of every branch.

We're often so accustomed to seeing things in a certain way that we become blind to the possibility of something we can't yet imagine. Often the way to create something wildly different is to step back and look at what stories we've taken for absolute truth, when really they're ripe for revision. What-If Framing often begins with looking at trends beginning to take shape, predicting how they'll play out ten, twenty, thirty years down the road, and asking how you might respond to them creatively. Set aside some time each week (or encourage your employees to do the same) when you think about the impossible. The point of this exercise is not always to come up with new ideas so much as it is to recognize the strength of your beliefs about the way things are. You might be surprised at how tightly you hold on to ideas without really knowing why.

While it is wise to follow Lisa Solomon's advice to take the long view when it comes to forecasting trends—what better way for people starting out in their careers to improve their job security than by looking at what jobs will be in demand in the decades to come?—it can be equally inspiring to look at the ways that industries like health care and education are already being transformed. In a piece for the *New York Times Opinionator* blog, Tina Rosenberg wrote about a Los Angeles–based organization called EngAGE, which brings classes in the arts to five thousand residents of senior housing. "We live in a society that's very acute-care based—we wait till someone's sick," founder Tim Carpenter was quoted as saying. "We decided to try to get people to take on healthy behaviors without having to go to the doctor." Carpenter reframed an industry by focusing not on illness but on keeping people well by giving them a chance to create things they might never have imagined they could.

At its most basic level, What-If Framing requires us to look at the world around us—our lives, our jobs, our stuff—and ask, "Why are things the way they are?" and "What if things were different?"

Playing

IT TOOK SEVERAL HOURS, BUT Harry West and his team eventually reached a conclusion about their current challenge: Drinking was weird.

West, the CEO of the Boston-based consultancy Continuum, had brought together a diverse group of his top people—collectively, they had degrees in packaging, design, business, engineering, human factors, and technology policy—to help redesign one of the greatest innovations in Swedish commercial history: the tetrahedron-shaped Tetra Paks now so common in Europe, Asia, and much of the world. Dr. Ruben Rausing is usually credited with the idea for the coated flat cardboard packages—picture a pourable pyramid—which haven't changed much since the first Tetra Paks came out in the fifties. They were designed for the way people drank—sitting down.

But at a certain point, it became clear to Tetra's leadership that people weren't drinking that way anymore. They were mobile, on the go, and wanted their liquid refreshment as they moved about.

After years of trying to fix the problem on their own, Tetra Pak's executives contacted West for help. Continuum then launched a major ethnography research project and collected tons of information about cultural drinking habits in Shanghai and Hangzhou in China, Milan and Modena in Italy, and Boston, the consultancy's home base. The data was rich, but data alone couldn't tell them what kind of product to design.

So West gathered together a team that had worked together before and trusted one another to be, well, a little nutty. They met in the firm's project room and started . . . drinking. "We were handing around bottles and packages, drinking from them, talking about how we felt about them, and we began to reflect on what we were doing, to think about it more carefully," West says. "It is a kind of an intimate experience handing a package back and forth with someone you work with, taking a sip and talking about how you feel. This is the sort of thing you normally only do with your spouse or lover."

As West and his team observed each other, they began to evolve a language for the differences in drinking habits—and the weirdness of it all. "We talked to each other about our preferred way to drink and expressed amazement and disdain for how other people on the team were doing it."

And that's when, West says, "It became fun."

FOR SOME TIME, AMERICAN SOCIETY has viewed play as kid stuff; it's been dismissed as trivial or marginalized as the territory of those lucky enough to work in creative fields or the arts. And there's some truth to the misconception. For centuries, musicians, painters, and dancers have utilized the strategies of play to create masterpieces. In a recent *Harvard Business Review* article, the sculptor Richard Serra, known for his huge installations of sheet metal bent into spirals, ellipses, and arcs, explained his process: "In play you don't foresee an end product. It allows you to suspend judgment. Often the solution to one problem sparks a possibility for another set of problems. . . . In the actual building of something you see connections you could not possibly have foreseen on that scale unless you were physically there."

Though there are countless ways of playing, play can be defined as tossing aside the rules of "regular life" for a period of time in order to follow new rules or try new possibilities. Play can exist within the structure of a formal game, but it doesn't have to. (In fact, the words "play" and "game" are interchangeable in a number of languages, including German, though we separate the two in English.)

We often aim to achieve a goal, but sometimes we play simply for the joy of it. Playing can involve strategies—some simple, some very complex. Some games teach you everything you need to know before you begin; in others, you learn to play as you play to win.

When we play, we try things on and try things out. We improvise, taking on new roles, imagining what would happen if we possessed new capabilities or behaved differently. We throw away what doesn't work and build on what does. We can play alone or compete against someone else; we can collaborate with another person or a team against a larger enemy. We may lose a game or a battle, but there is always the chance to start again.

PLAY HAS ALWAYS HAD ITS serious side. Though scholars are in disagreement over chess's beginnings, an early form of the game called *chaturanga* (a Sanskrit word that referred to the four divisions of warfare: infantry, cavalry, elephantry, and chariotry) was played in India around the fifth or sixth century before migrating to Persia and later to Europe, beginning with the Arab conquest of Spain. "This was a war game," writes David Shenk in his history of chess, *The Immortal Game*, "where ideas were more important and more powerful than luck or brute force" and where understanding was "the essential weapon."

According to military strategist and historian Max Boot, the author of *War Made New*, a precursor to the probability-based war-gaming scenarios we use today can be traced to the Prussian

General Staff between 1803 and 1809. Staff offices moved metal pieces around on a map or sand table, with blue pieces representing Prussian troops and red pieces the enemy. Dice were used to inject a note of chance to the battles and an umpire scored the game and decided the winner.

War games continue to be an essential part of US military strategy and training. In 2002, General Tommy Franks, top officer at Central Command, used the planning exercise Internal Look in preparation for the invasion of Iraq, successfully forecasting how the Iraqi army would fight based on its training and the tactics used in previous conflicts. Internal Look was also used in 2012 to assess possible outcomes of an Israeli strike against Iranian nuclear facilities. The outcome of this particular game showed significant US casualties if the Israelis launched a strike at Iran—two hundred American sailors dead—and led to the United States to caution Israel against such action.

Where military strategists have gone, economists and business thinkers have followed. The most prominent economic thinkers have used games as a metaphor for capitalism for over a century. In 1905, German economist Max Weber theorized that the secularization of a religious "calling" was what drove capitalism and gave the economic model the quality of a game. "In the United States," he wrote in *The Protestant Ethic and the Spirit of Capitalism*, "the pursuit of wealth, stripped of its religious and ethical meaning . . . actually give it the character of sport."

American economist Frank Knight took the metaphor further: "Industry and trade is a competitive game, in which men engage in part from the same motives as in other games or sports." When it came to the motivation of those who wanted to build great business empires, he wrote, it "is not a matter of want-satisfaction (money) in an direct or economic sense . . . but simply an insignia of success in the game, like the ribbons, medals and the like which are conferred in other sorts of contests."

For the past thirty years, Knight's insights have been lost to a US business culture enthralled with rational, efficient market theory. Yet among creative entrepreneurs, playing is a key discipline and strategy. When Barbara Walters asked Google founders Larry Page and Sergey Brin what they considered the most important factor behind their success, to her surprise, they didn't say it was their college professor parents or their Stanford University engineering degrees. "We both went to Montessori School," Page said, "and I think it was part of that training of . . . questioning what's going on in the world, doing things a bit differently." Montessori schools use games to teach children how to discover knowledge. It's perhaps no accident then that Montessori alums include Wikipedia founder Jimmy Wales, Amazon founder Jeff Bezos, Sims video game creator Will Wright, and rap mogul Sean "P. Diddy" Combs.

Other entrepreneurs with educational backgrounds in art, design, and music where play is intrinsic to learning have founded a whole slew of new companies, including Kickstarter, Tumblr, YouTube, Flickr, Instagram, Vimeo, Android, and, of course, Apple. And the list goes on and on: Paul Graham, the founder of Y Combinator, one of the top incubators for new start-ups in Silicon Valley, studied painting at Rhode Island School of Design and the Accademia di Belle Arti in Florence, in addition to getting his PhD in computer science from Harvard. Biz Stone, cofounder of Twitter and Xanga, says he learned a valuable lesson studying graphic design. "Being playful, less structured, less hierarchical . . . ," he said as an example of what he might offer to MBA students at the Berkeley Haas School of Business before he became an advisor there. Stone himself dropped out of college twice, but by inviting him in, Berkeley is demonstrating conviction that his playful approach is the way to build better businesses and organizations.

The desire to bring playfulness to work is not in and of itself new. During the first dot-com boom, it seemed that every other business story was about some hot new start-up whose offices looked like the inside of FAO Schwarz. Those of us in the business reporting and consulting worlds certainly embraced the idea that business could be cool and fun. "Innovation gyms" were built on corporate campuses all over the country, stuffed with beanbag chairs, moving tables, whiteboards, and Post-its. When we were developing a new magazine, *IN: Inside Innovation*, at *BusinessWeek*, I took over a small meeting room, took down from the walls the photos of heroic CEOs with their arms crossed in front of them (all male except for Carly Fiorina, then-CEO of HP), and put up pictures of green-faced children and burnt orange landscapes.

The idea was to generate creativity by having fun. In practice, many of us assumed that meant bright colors that were supposed to wake us up, inspire us, help our teams brainstorm a hundred new ideas in an hour, maybe a thousand. Big breakthroughs, what business schools and business managers now call "disruptive innovation," surely would emerge from all those fresh ideas. Many people in corporate America were playing this way . . . but did these techniques actually work? *BusinessWeek* successfully launched *IN* in 2006, but I would be hard-pressed to say it was because of anything our innovation gym generated. We rarely used the whiteboard and met mostly with the magazine designer in the art department. In fact, bringing Katie Andresen, from media firm Modernista, to the team was far more instrumental in ginning up our creativity than redesigning our space and spitballing ideas.

In retrospect, it was naive to think that just throwing people at each other, either deliberately or serendipitously, would generate creativity. Good teams require trust and skills and knowledge, not simply unfamiliarity and modular furniture. Colors are nice, and changing them is good, but . . . really. Besides, two members of my five-person team, including myself, were color blind. The play in those innovation gyms, in those years, was a bit silly.

In 2008, I went to a conference called "Serious Play" at the Art Center College of Design in Pasadena organized by Chee Pearlman, five-time winner of the National Magazine Award for her work as editor of ID magazine, a top design publication at the time. Presenters included Charles Elachi, then the head of the Jet Propulsion Laboratory, who called the JPL a "Disneyland for nerds and a playground for adults." Michael Curry, a master of puppetry, attributed his successful work on *The Lion King* to his studio, a "playground of tools and machines," where welders and seamstresses and sculptors and painters came together to create new nonlife forms. Tim Brown, President of IDEO, one of the top design and innovation consultancies, talked about what happens in play that makes it so special. "It is friendship that yields true play," said Brown, "allowing us to check our fear, our embarrassment, and defenses at the door."

A clear theme emerged in many of the discussions I had with leaders of some of the most creative organizations out there: simple silly play on its own doesn't lead to innovation. As the Continuum team discovered when seeking to come up with the new Tetra Paks, the best ideas emerged out of a process that involved a variety of players who trusted one another working together toward a specific goal.

Not long ago, Craig Wynett called me from P&G's Innovation Gym, which was supposed to generate hundreds if not thousands of new ideas for new products, sales, and profits. It hadn't. Wynett said the technique of throwing out new ideas out of context had not led to the breakthroughs they'd hoped for. P&G's Innovation Gym failed not because it wasn't a great playground for creativity (it was a big, beautiful space), but because they weren't playing the kind of games that helped them adapt to a changing environment and achieve new goals. As I wrote in chapter 3, P&G has had much more success with "Connect + Develop"—a strategy that opens up silos and connects people at P&G with scientists outside the company.

The play I'm talking about isn't about picking the right color for your offices or shouting out hundreds of ideas about things you don't know very much about. In serious plays there are rules, there is competition, there are winners and losers. Above all, there is learning, the kind of learning that allows you to navigate unknown areas, make unusual connections, and achieve new goals in unforeseen ways.

PLAYING, NOT PROBLEM-SOLVING

In times of relative stability, a conventional "identify the problem/find the solution" strategy is adequate. Problem-solving approaches work—but only when you know the problems. But today there are so many "unknown unknowns" (as former defense secretary Donald Rumsfeld put it), that we don't know the questions we should be asking, let alone the answers. In an uncertain, complex world of constant change, playfully discovering new answers to puzzles that do not have one right answer is a better approach than solving existing problems that do.

We need strategies that are dynamic and open-ended— strategies that are already being put to the test by some of the world's most creative individuals, though they'd never call it "strategy." To them, it's just "messing around." Chad Hurley, the cofounder of

YouTube, described the process for coming up with the company's name in an interview with the late Bill Moggridge, cofounder of IDEO and director of the Cooper-Hewitt National Design Museum. "We just did sketching on a whiteboard," he said. "It was just a play off the word "Boobtube." We were playing off that. Someone came up with the name, and I registered it."

The language of founders/entrepreneurs is the language of play. They're not out there solving problems; they're out there playing—and it's during this process that they're coming up with disruptive technologies and life-changing products.

Play triggers competition and cooperation, tenacity and joy. When people are playing, they take risks they would not ordinarily take. They experience failure not as a crushing blow but as an idea they tried that didn't work. Play transforms problems into challenges, serious into fun, one right answer into many possible outcomes. By not limiting yourself to one right answer you open yourself to contemplating outcomes you never might have imagined.

PLANNING FOR PLAY

Walter Isaacson's biography of Steve Jobs has fascinating detail about Jobs's interactions with people, but it's only in chapter 26, deep into the book, that Isaacson opens the door of Apple's center of creativity—the design studio—a crack. "This great room is the one place in the company where you can look around and see everything we have in the works," said Jonathan Ive. "When Steve comes in, he will sit at one of these tables. If we're working on a new iPhone, for example, he might grab a stool and start playing with different models and feeling them in his hands, remarking on which ones he likes best. Then he will graze by the other tables, just him and me, to see where all the other products are heading. . . . He gets to see things in relationship to each other, which is pretty hard to do in a big company. Looking at the models on these tables, he can see the future for the next three years."

Ive goes on to describe the studio as calm and gentle, "paradise if you're a visual person. . . . There are no formal design reviews, so there are no huge decision points. Instead, we can make the decisions fluid. Since we iterate every day and never have dumb-ass presentations, we don't run into major disagreements."

What Ive is describing sounds a lot like what Dutch cultural historian Johan Huizinga referred to as a "magic circle." In his 1938 book *Homo Ludens,* Huizinga analyzes the role of play in cultures around the world. "The stage, the screen, the tennis court, the court of justice, etc. are all in form and function play-grounds," wrote Huizinga. Magic circles are "temporary worlds within the ordinary world, dedicated to the performance of an act apart."

Building a special space away from normal activity, where people trust each other and agree to behave by a different set of rituals, is key to enhancing your team's creative capability. Creating magic circles enables a group to solve puzzles, connect dots, prototype, make mistakes, and learn from them. In other words, the magic circle is where the competencies of creativity get "played" out. Knowing how to build one is itself a creative skill.

Magic circles are ideal environments for innovating products like the new Tetra Pak or, as happens in the Apple studio, for tweaking designs without the need for a more formalized, bureaucratic review process. No two playgrounds should work exactly the same, and there's no one-size-fits-all formula: Individual team members will naturally bring different work styles and talents to the table.

What's important is that you set some parameters. Just as games like soccer and football need rules, so too does this kind of work. But you can't be so rigid that you limit the free flow of ideas: A magic circle requires a balance of structure and looseness. Setting up these parameters allows you to get into the zone together, interact smoothly, and quickly move toward a goal.

Choose people who trust each other enough to suspend judgment for a time. Trust is hugely important; people need to be willing to fall on their faces, make asses out of themselves, and learn from it. Remember, you're not solving a problem; there's no such thing as the "right answer." Very often the only team you need is just one partner, perhaps two. Don't confuse quantity for quality. It's less important to come up with a lot of ideas than it is to get the right people in the room and build upon your shared experiences to create something new.

And finally, don't be afraid to use your hands. Or—as the Continuum team discovered with their work on the Tetra Paks—your mouths. It was through this kind of hands-on play that they concluded there are three main styles of drinking: "Sucking," "Pulling," and "Pouring."

In everyday life, we don't typically look closely at people's mouths as they eat and drink. We tend to avert our gaze. But in that room, the Continuum team moved beyond the normal rules and restraints of society. They stared. They didn't hold back their own "weird" habits. They imitated each other, again and again, laughing and choking, until there was clarity.

The team's insights—that the way we drink is not only personal and cultural, but also depends on the drink itself—came from the free interplay of West and his team's "fooling around," and their willingness to embarrass themselves in front of each other in their project room. With their new vocabulary, Continuum went back to the field to test out designs that enabled the three kinds of drinking patterns. They even put cameras inside different packages to see exactly what people were doing with their mouths and lips when they drank. "We could now be very directed and purposeful in our creativity," says West.

The solution? "Hole over the edge"—a design with a larger hole for drinking than any previous Tetra Paks, with a laminated lip located right over the front folded edge of the package and a reclosable cap. That enabled all the Suckers, Pourers, and Pullers to drink easily and with pleasure, even while walking. It would take Tetra three years to work out the engineering to develop the Dreamcap Tetra Paks. They were introduced in 2011 for distribution in the Middle East and came to the United States in the summer of 2012 as packaging for Gatorade, Pepsi's sports drink.

IF YOU STILL THINK YOU don't have the capacity for Creative Intelligence, try reframing aspects of your work or your life in a more playful way: Approach obstacles at work or in your life not as problems to be solved, but as challenges to be met. When a new task comes your way, put together a team of people you trust, encourage each other to be honest (that's another area where serious play diverges from brainstorming, with its emphasis on positive encouragement at the expense of open dialogue), and let everyone know it's okay to look a little silly.

And your play doesn't need to end there. While fostering your own playful mind-sets and organizational cultures are essential for creativity, there's another aspect of play that is transforming the way we work and live: gaming. All around us, people are using the concepts of games to create new ways of engaging and improving their lives.

WHAT'S THE GAME?

After years of telling people how to live better by giving them feedback data on their weight, exercise, and diet and smoking habits, policy makers and health-care providers have discovered another strategy. Adam Bosworth, the founder of Keas, a "gami-fied" corporate wellness program in which employees team up and compete by meeting health goals, calls it "the Power of Play."

"If you want to get people to do something that they might not otherwise do—such as plan for retirement, study for exams, learn something new, get healthier or pick a school for their kids—make it a social game," wrote Bosworth in a 2011 *Tech-Crunch* post. "You will see engagement rates change from a mere 5% to 70% or more, and people will sustain those rates week after week after week. Why? Because games are fun and appeal to the primitive brain in all of us that wants constant rewards, social recognition and adventure."

When he headed the development of Google Health between 2006 and 2007, Bosworth wasn't thinking of the Power of Play. He and his team looked at the US health-care system, with its mess of different and conflicting interests—from insurance companies and hospitals to governments and patients—and decided, in true Google fashion, that the problem was lack of access to data. The Google Health team believed that they could alter personal health outcomes—help people lose weight or lower their cholesterol or blood pressure—by delivering personal health records to patients and medical records to doctors.

But it didn't work. Eric Bailey, Aimee Jungman, and Thomas Sutton highlighted some of the reasons why on Frog Design's site Design Mind: A major issue was that the information itself wasn't meaningful to people. The personal health records had tons of data but didn't provide enough of a narrative framework to help people understand the steps they needed to take to wellness. In addition, the Google Health technology didn't offer a platform for people to share information about their illnesses with family, friends, and, most important, other people with the same disease. Moreover, it was hard for the patient to share data with his or her doctor because personal health records weren't integrated with the physician's. And while the records were accessible on computers, you couldn't download the info onto your smartphone, which didn't help users who wanted access to their health information on the go.

But even assuming Google Health became more social or developed an easy-to-use app, a key ingredient was missing. Bosworth and his team believed that helping people see the discrepancy between the starting point and the goal would be enough to push people to change those numbers. It was logical. It was efficient. It was wrong. Google Health attracted far fewer users than expected and closed after just over three years.

What Google didn't get was the game.

BOSWORTH DIDN'T LET THE FAILURE at Google Health stop him. Three years later, Keas is one of over a dozen new wellness gaming programs being used by organizations dealing with the financial cost and loss of productivity that come with disease. With more than one third of US adults now officially obese, that cost is substantial and growing.

According to *Harvard Magazine*, more than 100,000 people from thirty-five companies have finished the twelve-week Keas program (corporations pay $12 per year per participating employee). Of the eight thousand employees of global construction corporation Bechtel who participated, roughly half said they lost weight. In twelve-week pilot programs with companies such as Quest Diagnostics, Pfizer, and Progress Software, Keas reports that 70 percent of employees stayed enlisted in its games throughout its duration and 30 to 40 percent posted every week on the game news feeds. People who reported weight loss shed an average of 5.5 pounds, and the proportion of people eating vegetables and fruits doubled from 37 percent to 73 percent. Half the employee participants said they were more physically active.

The big challenge for companies like Keas is to keep people playing at the game of health over the long term. But many companies are taking the chance that the game will continue.

Keas's attempt to gamify good health habits is just one of many examples of the way creative individuals and companies are using play and gaming to motivate people to make better decisions and make work more enjoyable. Virgin HealthMiles (Richard Branson is always a bellwether of what's new and at the edge) and RedBrick Health are two other ventures that offer well-being games to employees of big corporations and organizations.

What these programs share is a drive to harness the power of fun and competition in order to get things, often difficult things, accomplished. Someone once told a married friend of mine that it would be easy to get her husband to clean the bathroom: Just turn the bathroom mops and brushes into power tools and he'd forget the nasty nature of the task. Perhaps as the growing interest in playing games to help us lose weight or learn history is a testimony to how powerful a force it can be.

SuperBetter is one of the most creative additions to the genre; it was invented by Jane McGonigal, author of the bestselling book *Reality Is Broken*, and a leading advocate of games and their power to improve our lives. SuperBetter is designed to help players build up the resilience necessary to meet personal goals through the completion of a series of quests. You can customize your own quest or choose from a number of goals such as losing weight, learning to meditate, or recovering from an injury. Though a passionate believer in games' ability to help us better weather life's challenges, McGonigal warns against forced "gamification." "I don't think anybody should make games to try to motivate somebody to do something they don't want to do," she said in a *New York Times* article written by Bruce Feiler. "If the game is not about a goal you're intrinsically motivated by, it won't work." And so SuperBetter's brightly designed site offers players a number of choices at every level of the game about which goals to set, which quests to embark upon, what activities or friends to turn to when they need a boost.

The building of choice into the game echoes what Huizinga said about play seventy-five years ago: "Play is a voluntary activity. . . . By this quality of freedom alone, play marks itself off from the course of the natural processes."

A lack of freedom is the major reason why so many corporate off-sites fail and why we've come to resent the idea of "trust circles" and relay races. Real play is voluntary—who, after all, wants to be pressured into playing games with our supervisors only to be evaluated for our "openness" and bonding skills afterward? Games that are foisted upon people—no matter how much fun they might seem in theory—lack the essential ingredient of successful play: trust.

THE PLAYERS MAKE THE GAME

Like so many people, I open my e-mail in the morning looking forward to something special. A message from a friend. Breaking news from an online newspaper or website. A cool Twitter link from my digital posse. And a great deal in shirts—or some other article of clothing from Gilt, the high-end fashion platform that offers me a chance to get cool clothes at up to 60 percent off—but only if I play it right.

Shopping on Gilt is not a simple transaction. The site incorporates many elements of a great game: uncertainty, because I don't know what will be offered or whether I'll succeed at procuring what I want; speed, because you have to be fast; and competition, because the best stuff sells out quickly. There's an element of learning: I'm getting better at knowing when to look and how fast to jump. Outcomes are varied: I often pivot away from one kind of clothing to another, occasionally winding up on Gilt's sister site JetSetter to peruse bargain rates for fancy hotels in London and Mumbai. Because Gilt constantly changes its offerings, I'm always curious and willing to play.

It's all a bit of fantasy (I actually buy very little since I'm not quite in the high-income/wealth sweet spot of Gilt's target audience) but for the short time that I'm on the site, I participate in a story of high fashion and high-class global travel. It draws me in. And other people as well. Gilt has grown quickly since it was founded. It raised another $138 million from investors Goldman Sachs and Japan's Softbank in 2011, and in 2012 its main flash-sale online retail business turned a profit just five years after its launch.

As with many of the Creative Intelligence competencies, a generational shift is a major factor contributing to the explosion in the popularity of gaming. To members of Gen Y who grew up with social gaming—playing with friends and strangers online, learning by gaming—game dynamics are a way of understanding and navigating the world around them. It shouldn't surprise us that people who grew up with a gaming social structure would prefer it in their adult lives as well. And many of us who are older can see in games a way of learning and accomplishing that is more effective than memorizing and being coerced. We're all gamers now.

Games today are more complex and varied than ever before—and more popular. At its height, 80 million people played Farm-Ville. And millions more people join guilds to learn how to be blacksmiths or miners, go on quests, and defeat dragons and enemies as part of the online role-playing game World of Warcraft. But now that the game has broken out of the world of entertainment and the dynamics of gaming are being used to help people get healthier, learn new things, and have more fun while making purchases, the game is fast becoming the dominant social structure.

Borrowing elements of gaming can come in handy whether you're building a website to reach customers or trying to motivate yourself or your team to avoid procrastination as you finish up that big project. But what game do you play? Where do you begin?

With the players.

Bob Greenberg, CEO and Chief Creative Officer of ad firm R/GA, discovered this when his client Nike approached the firm for help doing something "social." Greenberg, a true knowledge miner, asked himself what runners did besides run. His first thought was: listen to music. What could be a more natural fit? Runners are always listening to music. So in 2006 R/GA hooked Nike up with Apple and linked Nike runners to their music with a wireless chip in their running shoes and an iPod. It also created a way for runners to store information on distances run or walked, calories burned, and workout times.

Initial success was modest. So Greenberg asked, what else do runners feel passionate about in addition to their music? And the answer was . . . competition. Running, like baseball and so many other sports, is driven by numbers. People are obsessed with their own times and the times of the people they run with and against. So R/GA designed a website where Nike runners could post all their data and also challenge people all over the world to beat them. The game worked. People wanted to play. Sure, running is good for you. But thinking about it in those terms isn't much fun. It's work. Racing—now, that's fun. The site has been a great success for Nike. It's created what business strategists are calling an "ecosystem," much like Apple's, that's composed of a community of consumers, followers really, participating in activities that are fun and meaningful to them. More than six million runners log on to see how they fared against their competitors.

Once you've thought a little bit about who'll be playing your game, you can begin to develop some rules.

BUILD YOUR OWN GAMES

There are many different kinds of games, but game designers frequently distinguish between two kinds: simple and complex. Simple games include puzzles and, like the *New York Times* crossword, they do not change as a consequence of the decision you make. Puzzles are simple to solve (though not necessarily simple to create, as anyone who's crafted a crossword will attest), they often are played alone, and they have exactly one solution.

A more sophisticated puzzle takes the player to a higher level of difficulty. Many puzzles become more difficult as you go along—people who can whiz through a *New York Times* crossword on Monday (the day the puzzle is easiest) might not answer more than a handful of questions on the dreaded Saturday puzzle. Whether easy or hard, simple games are part of a closed system; the player doesn't interact with other players.

Complex games, however, are a different story. They can begin with simple elements, and perhaps even stay that way, but the solution is always less cut-and-dried than with a puzzle. In poker, there is no fixed "right" answer. The dynamics of the game shift with every move and new information is constantly being introduced—both about the cards and about the players. Complex games are open systems; players receive information from each other and often from outside the game.

In other words, complex games mirror the volatility, uncertainty, and ambiguity of the real world, which is why understanding them—their rules and rituals—is a compelling strategy for meeting the challenges of our lives today.

Much like "flow," complex games can seem to exist within their own time and space, a world separate from ordinary life. They require people to take a break from their routines and try out a different set of behaviors, expectations, outcomes, and rituals. Players of complex games agree to a general set of rules and goals, but the strategies for meeting those goals are fluid and dynamic. The game begins when everyone decides to go along for the ride, even though the outcome may not be totally defined. The game ends (as every parent knows) when the players say it does.

But perhaps the most intriguing aspect of complex games is their ability to teach us as we play. Playing games, in fact, was key to the progressive education movement's method for teaching children. Friedrich Froebel, founder of Germany's kindergarten movement of the 1840s, designed wood cubes and blocks for children to learn by building; in so doing, he linked work and play as expressions of the same creative activities.

In the early twentieth century, the progressive education movement expanded, bringing its focus on play and games to years K–12 with the Montessori and Waldorf schools. These schools were developed at the height of the Industrial Age and their reliance on blocks, boards, rods, beads, pegs, and boxes reflects that era's sensibility. Their strengths lie, in part, in the simple physical materiality of the wooden "toys" and board games used in the play.

In order to adapt to an increasingly digital world, some schools have begun to embrace social gaming as well. In 2007, Katie Salen, a graduate of the Rhode Island School of Design and former director of the Center for Transformative Media at Parsons The New School for Design, received a MacArthur Foundation grant to set up Quest to Learn (Q2L), a public school that "gamifies" traditional grade 6–12 curricula.

Salen puts on a weeklong summer camp to train fifth grad-
ers in the skills of designing location-based games. At the end,
these children have improved the following competencies: digital literacy, creative problem solving, and collaboration. They've
designed and traded digital avatars via Bluetooth, solved myster-
ies with GPS tags, and created and played their own digital and
"physical" games.

While these kinds of summer camps for adults don't exist
(yet), what follows are some basic principles of game design that
you can keep in mind when you're embarking on a new project,
meeting with a team, or planning to deal with a serious issue that
has resisted resolution so far (like getting kids to eat vegetables).
Playing a game may provide the motivation and rewards to get
people to change their behavior and their goals. Perhaps that is
why 72 percent of US households play digital games.

1. Complex games are dynamic and require adaptability.
StarCraft II, one of the best-selling online games in 2010, has
so many players around the world that a championship was held
in Providence, Rhode Island, offering a grand prize of $50,000.
The game appears to be simple—each player picks one of three
species, terran (humans), zerg (insectoid creatures), or protoss
(photosynthetic aliens); then they battle for control of a territory.
But every decision brings in a host of new variables and so players
must continually switch up their strategies.

This game's dynamics mirror life in any organization today,
be it a large corporation, a new start-up, or an art gallery. It reflects our personal lives as well. Look at the typical day of a working parent and you'll see the busyness and constant change-ups
that mirror the decision-making skills necessary to win StarCraft
II. Playing dynamic games can be good preparation for making
solid decisions even when you're juggling a lot.

2. Complex games depend on possibility, not probability.
There are few things more terrifying to both children and parents than cancer. Somehow, we feel it's a disease that should never strike the very young. To children, already struggling to gain some control over their lives, cancer makes them feel even more helpless. Re-Mission is a game that teaches teenagers with cancers such as leukemia, lymphomas, and sarcomas how their treatments wage battle on cancerous cells. The game stars Roxxie, a "fully armed nano-bot" who leads players through "challenging missions and rapid-fire assaults on malignant cells, wherever they hide."

The game was created by HopeLab, a company founded by Pam Omidyar (wife of eBay founder Pierre Omidyar), both a scientist and game lover who began to wonder "if giving young cancer patients a chance to blast their cancer in a video game might actually improve their health." According to a study conducted by HopeLab that was published in *Pediatrics*, the game "significantly improved treatment adherence."

Because they exist outside "real life," games introduce the notion of possibility into even seemingly impossible situations.

3. Complex games use "scaffolding."
The best games introduce just enough new information so that you can pass a level or make a particular decision, but not so much that it slows the game down. You may encounter trials where you must try out a particular tool, weapon, or skill, all of which will prepare you for a bigger challenge later on. "A game designer thinks about what a good teacher thinks about," says Salen. The designer asks, as would a teacher, "What is it that my player needs to know at this moment to learn to meet the challenge and achieve the goal?"

If you are one of the many millions who play Angry Birds, you know how scaffolding works—literally. Scaffolds built by oinking pigs get increasingly more complicated and difficult to knock down, but luckily the little birds you catapult at the pigs are equipped with more sophisticated bombs or flying patterns so that you can better meet the new challenges. The subtle techniques you learn in the first few levels of each new "world" prepare you for more complicated levels down the line. In other words, you learn as you go (and go a little crazy in the process).

In SimCity, one of the most popular games in the world, building your own city gets increasingly more complex and difficult as you proceed. At the beginning, you can ignore most of the aspects of the simulation as you begin to build your city. But gradually, you have no choice but to learn about limiting pollution, reducing crime, managing tax rates, removing waste, and all the other functions that go into making a city viable.

One of the best ways to prevent ourselves from creating something new is to say, "I don't know how to do it." But most complex games don't come with rule books; you need to learn by doing. When you're working on a project, take notice of new skills or insights you've picked up along the way; they will surely come in handy when you're tackling bigger challenges down the line. And continually take the time to check in what you've already learned—in your industry, in school, in life: You probably know more than you're giving yourself credit for.

4. Complex games often look simple.
Humans vs. Mosquitoes is a game that has no equipment, complex rules, or computers. It can be played outside in a field or on a tabletop. It was designed by students at Yale and Parsons to help children understand how to eradicate dengue fever disease in Africa, Asia, and Latin America. Like malaria, the fever is spread by mosquitoes; the goal of the game is to remove all mosquito eggs from the breeding grounds before mosquitoes kill the humans.

Players are split into two teams—mosquitoes and humans.

Each mosquito gets a number of eggs (bottle caps or stones will do) to create a breeding ground while the mosquitoes try to bite the humans, and each human gets a number of blood tokens for protection while the humans try to clean up the breeding grounds. With every turn, players must choose between a defensive and an offensive move. If the mosquitoes bite the humans before they clear all the breeding groups, they win. If not, the humans win. It takes about fifteen to thirty minutes to play.

Though Humans vs. Mosquitoes may appear to be simple, it proves that you don't need a ton of resources to create complex games that teach valuable lessons.

5. When you can't win the game, change the rules.

In 1485, Leonardo da Vinci sketched the first human-powered airplane on record, an aircraft with birdlike wings. It was not only a beautiful work of art but also something of a game: Though several people had risen to the challenge of making the sketch a reality over the years, in 1959 a wealthy British businessman, Henry Kremer, formalized the creation of a human-powered aircraft into a contest. According to *Fast Company*'s Aza Raskin, Kremer offered a huge sum, 50,000 pounds ($1.3 million today) to the first person to fly a human-powered aircraft in a figure-eight around two markers situated half a mile apart. He would also give 100,000 pounds (about $2.5 million today) to the first person to make the trip across the English Channel.

People set about solving the problem, building planes that, according to Raskin, never even survived their test flights or that could go only a few hundred meters before pilots would be forced to land, too exhausted to go any farther.

Nearly two decades passed with a bit of progress, but no breakthrough. But that's the nature of progress, right? It takes time. You play the game by the rules—in this case, the rules were to build a prototype, test it, and then build another one—and you work toward results.

Aeronautics engineer Paul B. MacCready, however, saw that the game itself was flawed. According to Raskin, the other contestants' trial-and-error methods weren't getting anyone much closer to winning, so MacCready developed a new set of rules. The game was designed not to produce a winning plane model, but to speed up the learning process: Instead of building and rebuilding a plane in a year, MacCready thought, why not do it in hours? Instead of using permanent materials and finished parts, why not prototype using Mylar, aluminum tubing, wire, and tape? Instead of trying to get it right the first time, why not, as Raskin writes, "find a faster way to fail, recover, and try again?"

MacCready was able to fly three or four planes in a day because he was able to fix and refix them. The cycle of building, testing, learning, and rebuilding went from months and years to hours and days. MacCready made a new game—one that was faster, more dynamic. A game that allowed him to learn faster, make decisions faster, and veer off in new directions faster. Six months after MacCready changed the game, he won.

PLAYING: STRATEGIES FOR THRIVING AMID UNCERTAINTY

Play is a complex behavior, but it's something we can all learn to do, or relearn. As children, we made up games, transforming our backyards, bedrooms, and streets into playgrounds. We chose who we wanted to play with (and who we didn't). And we set rules. Walk into many science or engineering labs, hot design studios, or product development departments and you will find people playing the same way.

So how do you bring more play into your life? First, it makes sense to recognize that you may be a member of a magic circle without realizing it. Book clubs, for example, have much in common with the kinds of playgrounds Huizinga described long ago: They require a team of people you enjoy spending time with, a decision to take a kind of journey into the unknown together, and trust—you want a group who can commit to sharing their unfiltered opinions, listening to and building on one another's ideas in order to reach insights none of you may have ever discovered on your own.

Our book clubs, our fantasy football leagues, our multiplayer gaming campaigns are all playgrounds; we might not view them as such, but they are in fact places of discovery for adults. What these activities also demonstrate is that people want to pick their own teams. Sometimes, in work situations, we're lucky enough to be matched up with the right people, but all too often innovation is stopped in its tracks when people are grouped together arbitrarily; their individual talents are never given a chance to shine for any number of reasons: Their backgrounds are too similar and so there's no real discovery; roles are not clearly defined up top; people do not feel comfortable enough to be honest about their own ideas or constructively critique each other's.

In building your individual creative capacity, one of the most important steps is to go outside yourself and find the right partner or team. When I talk to entrepreneurs, our conversations return again and again to what they did with their buddies. Very few tell me about being alone and coming up with the big idea. It happens, sure, but more often than not this kind of discovery emerges as part of a social relationship.

This is what the four founding members of the Upright Citizens Brigade, a comedy troupe with an enormous cult following, understood when they left Chicago to bring their signature style of improv and sketch to New York in 1996. In a 2011 interview with the UCB Theatre's Artistic Associate John Frusciante and Academic Supervisor Will Hines, founding member Ian Roberts explained that in their early days, the UCB Four (Roberts, Amy Poehler, Matt Walsh, and Matt Besser) made commitments not to take any jobs that would take them away from the troupe. This ran counter to the way sketch groups were typically created: Theaters would cherry-pick performers and cast them on their own house teams. "Our feeling, and I still feel, (was) that the best sketch shows are an ensemble that chose to be together," said Roberts. The UCB's strategy paid off: The four comedians created a TV show for Comedy Central that has become a cult classic, and launched two theaters in New York and one in LA, as well as an acclaimed training program that has groomed some of the biggest names in comedy today. UCB alums and performers include *Anchorman* director Adam McKay, former SNL cast member Horatio Sanz, *Hangover* star Ed Helms, and a number of up-and-comers like Donald Glover, Ellie Kemper, and Bobby Moynihan.

Whether you're embarking on a "solo" creative project (a book, an album, a blog) and are looking for some feedback, or launching a larger endeavor that requires partners, often the fastest way to jump-start your creativity is to find the right teammates. Ask yourself who you "play with" at home and at work: Do you need more people in your life who spark your creativity? What skills do you lack that another person might excel in: Do you need, for example, someone who's good at finding sources of funding or who excels at logistics to help implement your ideas?

As much as we love stories about "serendipity," seeking out the right people may be as important as, if not more important than, accidentally bumping into them. It's very hard to play with strangers that you can't trust.

So how can organizations give people more control over their teams? Too few corporations hire with teams in mind. They tend to hire the smartest or "best" individual for the job without much thought about how that person will engage with others. Yet almost all the tech and media start-ups that I know that are growing fast do something different. They test applicants for how they work in ensemble. Since most creativity is social and doesn't come out of one individual's effort, this is a strategy that larger companies should at least explore.

As more companies outsource innovation to India and China and other places that have cheaper scientists and engineers, they are effectively destroying their in-house centers of creativity. Corporations who've employed "24-7 teams" strategies should give serious thought to rebuilding their labs so that innovation is more organic. There are plenty of reasons to have global laboratories, but creating them to save money is not one of them. I suspect that the creation of innovation gyms was a sincere attempt to revive the corporate lab. But an innovation gym without the full capabilities of a corporate lab is a little like a fitness gym where people talk about good workouts instead of taking classes and training with skilled professionals. It's one thing to get together and talk about innovation, but you need scientific knowledge and equipment to actually make anything happen.

Serious play turns the process of play into an instrument of change. Serious gaming does the same thing. Games can help us explore our limits and push beyond them. The next time you're facing a challenge at work, consider reframing it as a game. If you've been putting aside some task because it's something you've never done before, keep in mind that many complex games do not come with a rule book—and yet we play anyway, learning the game as we go along. The same is true of many of life's great challenges. Writers don't know how to write a novel until they sit down and do it; entrepreneurs don't know how to launch a company until they take the leap. Sure, there are countless books that teach craft and theory, management and finance, but until you've actually done the work, you never really know what you're getting yourself into. Games can be a great training ground for jumping into challenges before we feel that we're ready, trusting ourselves to learn as we go.

For organizations looking to bring elements of a game into their culture or their offerings, it all comes back to the players. As Greenberg learned when designing Nike's new site for runners, the games that work are ones that build upon what the players want. You can design the best-looking, most sophisticated shooter game in the world (such as the hugely popular video game Call of Duty), but if your target audience doesn't like fantasy violence, your game is over before it's even begun.

Making

IT WAS 3:44 IN THE morning on May 22, 2012, when the huge multistage rocket lifted off from Cape Canaveral, Florida. Bright white against the black sky, spewing clouds of light, the Falcon 9 moved steadily up, up, into the sky. Just minutes later a mission commentator announced that the rocket had passed the first threshold for potential disaster—the main stage successfully separated from the second stage. After a few more minutes, the Dragon capsule riding on top of the rocket successfully separated from the second stage—and the second possible crisis was averted. The Dragon capsule was free, but blind—no power. As the solar panels that provided energy to the capsule unfolded, the Dragon successfully made it past the final moment of danger.

Just days after the launch, the Dragon capsule, containing one thousand pounds of supplies, approached the International Space Station, flying in orbit around the earth. The Dragon made contact, and the station extended a huge arm that locked onto the capsule and brought it in safely. In a few hours, astronauts would unload the cargo from the Dragon capsule, rocketing the business of private space cargo transport into reality.

As someone who grew up during the *Sputnik* era, I still get a chill when I see a rocket heading into space. I wanted to be an astronaut even before I wanted to be a fireman, so I watched nearly every takeoff in the sixties and can still hear the NASA voices describing the gigantic Saturn rockets blasting away to the moon. When the Shuttle came along in 1970s, I followed those liftoffs as well. Space flight has always seemed to me the ultimate challenge: to be free of the planet and exploring the universe. Wow.

I replayed the video of the Falcon 9's takeoff from the same Cape Canaveral Air Force Station many times, and each time the voices of the young man and woman announcing the launch jumped out at me. They sounded younger to me than the ex-military NASA voices I'd heard as a kid and, while there have been a number of pioneering female astronauts, the "official" voice of American space flight had always seemed to me to be male. It was truly exciting to hear, and reminded me how much space travel had changed since its early days. This flight was, after all, not a NASA voyage, but the maiden trip of a company launched by a dot-com billionaire.

Many know Elon Musk as the co-founder of PayPal, the electronic system that allows people to pay and transfer money in P2P, or person-to-person, transactions, which have become the backbone of nearly all Web commerce. Without the company, we would be sending checks, money orders, and maybe even cash to eBay and Amazon every time we bought something online. PayPal got its start pretty much at the nadir of the dot-com bust, when Musk combined his company, X.com, with another e-commerce site, Confinity. By 2002, eBay realized that half its customers were using PayPal, so it bought Musk's company for $1.5 billion.

Many entrepreneurs have enjoyed phenomenal success in the shift to a digital economy. With a company that made e-commerce possible, Musk has been a standout. But instead of doing what so many successful Silicon Valley entrepreneurs have done before—become a venture capitalist and invest in other digital start-ups—Musk took a different path.

In 2002, Musk became the CEO and Chief Designer of the

Space Exploration Technologies Corporation, or SpaceX, a private commercial company that designed and manufactured in its own California factory the Falcon 9, the rocket that would replace the NASA shuttle in supplying the International Space Station.

Musk didn't stop there. A year later he founded a second company, Tesla Motors, to make cars. This being the twenty-first century, Musk decided to build all-electric cars and set up shop not in Detroit, but in Palo Alto, near Stanford University. The first two-door Roadsters came off the assembly lines in 2007—for a mere $109,000 each.

You would think that Musk would consider successfully launching the first private commercial rocket to dock with the International Space Station to be his greatest accomplishment. But Musk said in his celebratory comments that what really got him excited that week was the "S" passing its last safety checks before going into production. Finally, after five years of preparation, the first mass-produced, electric-powered four-door sedan—costing "only" $50,000—would be cleared for assembly in an old California factory.

Perhaps Elon Musk's journey signaled a shift in the culture of America—a shift back to the making of things.

A MAKING RENAISSANCE

In 2007, just before the biggest financial crash since the Depression began, 41 percent of all US corporate profits went to Wall Street investment banks, commercial banks, and other financial institutions. This was astonishing. Historically, banks never accounted for even half that figure. And rightly so. Banks were supposed to support business and keep the "real" economy going. But by the first decade of the twenty-first century, finance was no longer just the fuel powering investments but also the motor, the gear shifts, and the steering wheel of the entire economy.

And so it was only natural that many of our best and brightest flocked to the industry where the action—and the money—was. In the 1960s and 70s, the majority of business school graduates joined large corporations in retailing, manufacturing, and other industries that were the bread and butter of the US economy. But by the end of the century, two thirds of grads from the top B-schools at Harvard, Stanford, Columbia, Wharton, and Chicago were going into finance and consulting.

Top bankers received astonishing salaries—the brightest grads from Ivy League universities saw salaries starting at $200,000 to $300,000 plus a fat annual bonus several times that—to use their understanding of mathematical models to create esoteric financial instruments that sliced and diced mortgages and other kinds of debt. So complex were the models created by the Wizards of Wall Street that few bank CEOs actually knew what they really were or, more important, how much they were worth. We all know how that turned out.

It wasn't supposed to be like this. Placing our trust in the experts in finance and trading was, we were all told, a crucial part of the New Economy. When *BusinessWeek* ran a cover on the New Economy in 2000, it was a celebration of technology, finance, strategy, consulting, sales, service, and experience. The message was that we should ship all our manufacturing overseas and concentrate on higher-level, value-added information activities. Using your head, not your hands, was considered a higher evolutionary state of economic—and social—affairs. It paid more and you "didn't have to get your hands dirty." Sure, jobs were lost as factories closed in Ohio and Michigan, but the service economy was expanding to absorb everyone.

There were early signs of protest against the rise of globalization; unions spoke out against the closing of their factories, human rights activists began to voice concerns about the dark side of cheap labor. There were marches and rallies in the United States and beyond. A huge, and violent, protest against free trade in Seattle. But, while a few companies were early adopters when it came to embracing sustainable practices and local manufacturing, outsourcing remained the dominant strategy.

A more widespread questioning came when white-collar jobs began to be outsourced. First engineering and software-writing jobs went to India, but they were soon followed by jobs in accounting, law, architecture, design. Hundreds and hundreds of thousands of service jobs in fields like IT and finance were shifted overseas to places like Bangalore and Manila. Only when both the manufacturing sector and the service sector began to bleed jobs did blue- and white-collar interests find common cause.

As Americans began to feel the effects of the Great Recession, it became clear that the economic benefits of the New Economy disproportionately went to a tiny elite, while the vast middle class saw immiseration and downward mobility. We witnessed an inequality gap that hadn't been as wide since the 1920s and the Great Depression. The Occupy Wall Street movement, with its rallying cry "We are the 99 percent," crystallized the sense that something needed to change. You knew something fundamental was about to change when both the Tea Party and Occupy Wall Street found a common enemy, publicly blaming "Crony Capitalism" for destroying the American Dream.

Even while the New Economy was at its peak, alternative ways of thinking and doing had begun springing up around the nation. Alice Waters's groundbreaking organic restaurant Chez Panisse served as inspiration for a local food movement, which, four decades later, has gone truly global. The restaurant's model, one in which all the ingredients are grown within a few miles of its Berkeley location, is now being replicated by thousands of restaurants in Chicago, Boston, both Portlands, Cincinnati, Los Angeles, and across the country. Craft became popular again, or perhaps it never really died, as the popularity of online e-commerce companies like Etsy confirmed.

Just as important, Gen Y began to use the digital tools offered by companies like Apple to make a huge variety of things, from hand-bound books to independently produced Web series. These tools are so easy to learn and master that most people still don't think of themselves as "creating" or "making" when using them—but that's exactly what they're doing.

By the end of the first decade of the aughts, these 2-D digital tools were joined by a whole series of equally inexpensive (if not yet quite as easy to use) 3-D printing tools that can print out different kinds of objects—from hinges and doorknobs to jewelry and toys—in a variety of metals and plastics. You can pay about a hundred bucks for a monthly membership to TechShop in a number of cities, and spend your evenings and weekends taking classes in anything from sewing and woodcutting to machining and 3-D printing.

The new "maker movement" is manifesting itself in so many different ways that it's easy to miss the overall trend. Many of us learned about the DIY movement from home shows that taught us that we, too, could build our own furniture, restore our own homes, sew our own fashionable clothing. But traditional art forms, too, are experiencing a true renaissance. Weaving especially is seeing a huge revival, with prices for rugs and baskets soaring in value at national art contests and at the annual Indian Market in Santa Fe and the Heard Market in Phoenix.

Make magazine, launched in 2005 with the goal of providing online tools to a community of people interested in creating things, sponsored its first Maker Faire in 2006. The Faires are modeled after old-fashioned county fairs, but instead of the best bulls or butter statues, the best projects—from algae-eating fuel pumps to fire-breathing robots—are rewarded. The Faires celebrate "arts, crafts, engineering, science projects and the Do-It-Yourself mindset," according to the magazine. Today they are held in cities all over the United States, and tens of thousands of families go to show each other what the kids (and moms and dads) have created by themselves.

On the digital side, the language of hacking has taken on new meaning—it's no longer just about breaking into computer systems, stealing information, or planting viruses. To many people, hacking means tinkering, improving. Instagram users play with filters to "hack" their photos, to make them different and more interesting (and to share them as well). There are even "bio-hackers" who buy machines to copy DNA and attempt to hack them into new life-forms.

A major driving force of the new maker culture is demographic. To boomers who grew up during the Cold War, with half the world closed off to them behind the Iron Curtain, the fall of the Berlin Wall and the opening of Eastern Europe and China held so much promise. Globalization was exciting for consumers eager for a taste of the wider world, and for corporations who saw an expanding global marketplace as a strategy for growth.

Generation Y, on the other hand, has seen the negative effects of globalization—the uncertainty of throwing in your lot with a big corporation that has no loyalty to you, the way outsourcing has made it harder for them to get jobs and advance careers, the inequality of the haves and have-nots—and they have begun to innovate a different path.

For Gen Y, participation plays a key role in their culture and their happiness. Boomers may have been happy to sit in front of their sets and receive new information and entertainment, but their children and grandchildren want to be the ones making it. They may watch TV, but they DVR it to watch what they want when they want. Most spend more time "on" their computer (the language is important here) than "in front" of their TV screens because they are actively doing something—making little movies for YouTube, connecting with friends, taking photos and uploading them to their Flickr or Instagram pages. They are active, not passive, participants in media.

They are also a generation in search of authenticity. Bombarded with as many as five thousand commercial ads a day and growing up in a digital, virtual world, Gen Y places a high value on the "real." So there is a huge revival in vinyl records because the sound is richer and deeper than what comes out of iTunes. In 2011, sales of vinyl albums reached nearly 4 million annually in the United States, up from 300,000 in 1993. Artists from Radiohead to Adele, Bon Iver, and the Black Keys are releasing new albums in the format, and classics from Nirvana and Metallica are selling better than ever. The best-selling record for three years straight? That's right: the Beatles' *Abbey Road*.

The generations who grew up in a fast-food world are craving authentic, organic meat and vegetables from nearby farmers. Knowing the people who grew or hunted for your food, who crafted your cabinet out of "found materials," is cool—but doing it yourself is even cooler. Today artisanal beers and vodkas and organic food from nearby farms capture the imagination. So does starting your own enterprise rather than working for a big global corporation. Of course, what's "cool" among the young eventually makes its way into the mainstream, and so we're seeing people of all ages joining in the fun, and making money while they're at it.

WHILE MASTERY OF THE PREVIOUS Creative Intelligence competencies—Knowledge Mining, Framing, and Playing—can help us boost our creative capacities, Making involves learning the tools that can help us bring that creativity to life—in our personal lives, the businesses for which we work, and the organizations to which we belong.

To anyone under thirty, there is no need to make the case that Making is a necessary part of Creative Intelligence. For Gen Y, making things online—whether fashioning their own Tumblr pages or creating entire virtual worlds in online games like Second Life—is practically second nature. They use Google and Wikipedia as Swiss army knives, accessing free information, reassembling and remixing it, and re-blogging it.

Those of us who are immigrants to maker culture may need a bit more convincing as to why these skills are important. Making has always played an important role in the creative process but, until recently, expensive technology and limited access to tools and small-scale manufacturing made it impossible for the average American to prototype ideas for new products or bring them to a wider audience.

But now all that's changing. My own creative process has been drastically transformed by computer technology. When I was writing on typewriters and carbon paper, editing the first draft was slow, so I did only one or two passes—compared with dozens or more today. Making copies was a messy process, so I got feedback from one or two people. Today I might write a blog post that reaches thousands and get good feedback from hundreds. That having been said, typewriters, like vinyl records, are making a big comeback within the Gen Y set. A new generation of writers is trying to slow down their process, think more, and feel and hear the physicality of their work as keys strike the paper. You can go to Etsy, Fab.com, or the Vintage Typewriter Shoppe and find a refurbished vintage Royal typewriter for $340 or less.

In the same way, people are using film for photographs again. Polaroid cameras are hot items; a couple of years ago the iconic brand announced it would discontinue some of its instant cameras, only to reissue them in limited batches shortly after, taking advantage of the analog craze. There are even folks eager to develop their own film, empowered by resources online with instructions for everything from assembling the darkroom to mixing the chemical solutions. The richness of the color, the patience necessary to get that one perfect shot are appealing in a digital age where most people shoot hundreds of pictures and then do a ton of editing in post-production.

Walk into any Apple store anywhere in the United States and you'll find middle-aged and elderly men and women learning digital tools in order to make photo albums of their kids or grandkids or create family Facebook pages. Whether by creating an object we can hold in our hands or launching a platform that allows us to share an idea with our community and beyond, diving into the Making stage of the creative process is hugely rewarding.

No matter your level of participation, making things expands our universe, enriching our personal lives and the lives of those around us. What better way to build our creative confidence than by turning an idea into something tangible?

HOMEGROWN IS THE NEW COOL

Jennie Dundas and Alexis Miesen are an unlikely pair of ice cream shop owners. By the time the two came together in 2007 to start Blue Marble Ice Cream, Dundas had been acting on the stage for years and Miesen had spent ten years in international development, including a stint as the first Field Director of the WorldTeach program in the Marshall Islands in 2002 to 2003, where she put together a strategic plan for local teacher development and improving school performance.

Their Brooklyn-based company is atypical in a number of ways. Blue Marble is an independently owned organic ice cream brand, a rarity in the East since Ben & Jerry's was bought by Unilever. The product is made with milk from grass-fed cows from nearby New York and Pennsylvania farms. All its ingredients are sourced as locally and "conscientiously" as possible. Launched in 2007, just as the Great Recession was beginning, Blue Marble has two shops in Brooklyn and sells its organic ice cream in stores throughout the borough and parts of Manhattan as well. It's scaling fast, even as it fiercely retains its local identity.

Miesen says the name Blue Marble came from the nickname given to the earth by members of NASA after they saw the iconic December 7, 1972, *Apollo 17* photo of the planet from space. Miesen and Dundas extended the "marble" metaphor to the look of their offerings: regular-size scoops are called "marbles" and the smaller ones are called "mini-marbles."

The name illustrates connection to the earth, to playfulness, and more important, captures the company's focus on what it calls "conscientious consumerism." These days, says Miesen, more and more consumers "want to know the origins of their food purchases and are recognizing the impact those origins have not only on the quality of the products but also on the well-being of our global community." Blue Marble's owners love when customers ask where they source their dairy, coffee, vanilla, cocoa, and other ingredients because it's a way for them "to connect with the faces, cultures, landscapes, resources behind what we're eating," says Miesen. The stores use only biodegradable cups, bowls, and spoons, which are composted.

Like Method, Blue Marble doesn't wear its commitment to sustainability as a hair shirt. Instead, it embraces its local sourcing and job creation as core business values. In the nineties and aughts, a new generation of philanthropies such as the Acumen Fund pioneered a new model called "social enterprise." Instead of handing out money to government institutions in Africa and Asia to help the poor with their inadequate health and education, capital would be invested in private businesses and entrepreneurs in these areas. They would hire locally, buy their goods from nearby farms and factories, and provide needed services. Though developed as a way to help the poor overseas, it is also the model for Blue Marble, Method, and a fast-growing number of start-ups in the United States. For a new generation of entrepreneurs, all enterprise is social enterprise.

Coming full circle to Miesen's original area of interest, Blue Marble rolled out a nonprofit venture in partnership with Eric Demby, cofounder of the Brooklyn Flea, an open-air market for vintage goods and local food vendors. Blue Marble Dreams quickly raised $80,000 in grants and donations to finance an ice cream business in Rwanda. Miesen was inspired by the idea after meeting Rwandan drummer and playwright Odile Gakire Katese at a theater workshop in Utah. Katese said that Rwanda is a land of milk, and that she thought the women in her drumming cooperative could make and sell ice cream as a way to make money, and, perhaps of equal importance, bring joy to the community. After raising the initial funds, Miesen and Dundas traveled to Butare to assist with training. The ice cream parlor is called Inzozi Nziza ("Sweet Dreams") and employs eleven women, supporting seventy family members. The Butare parlor, Rwanda's first, sources ingredients from dozens of local dairy farmers, coffee growers, and beekeepers. Blue Marble has, in other words, exported its "local" model.

For many Americans, it's no longer enough to "buy it local" when the "make-it-local" culture has never been easier. Babette, a successful women's fashion line, makes all its clothes in a factory in Oakland (using high-quality textiles made mostly in Japan). The hippest hipsters in the Williamsburg neighborhood of Brooklyn butcher their own pigs from nearby farms and cook them at home with family and friends. And the most difficult restaurant in the world to get a reservation for is Noma, a Nordic "foraging" restaurant in Copenhagen that uses only local foods, gathered and prepared to accentuate their relationship to each other and their source.

A homegrown economy—one consisting of partnerships and mutual support between friends, families, and neighbors—is becoming increasingly meaningful to many people. There is a spreading sensibility, particularly among young people, about the importance of cultivating your own space and community. The "local" phenomenon in America cuts across political, class, and regional lines. It's an idea that could well unite conservatives and liberals, the Tea Party and Occupy.

"Local" is all about making things within your community with people you trust who share your values. It's a metaphor for generating jobs in the community and, with them, income and a better life. It also carries with it an implicit promise of sustainability—by cutting the energy used in importing from China and other distant locations.

Making and crafting things personally and locally have prestige and status once again. The phrase "Made in Brooklyn" has real meaning today, as does "Imported from Detroit." The movement may have started with locavore food culture—growing your own food in gardens in Ohio, buying food from local farms in New York, eating at slow-food restaurants in San Francisco. But local has gone, well, national.

LOCAL GOES GLOBAL

It's not just start-ups that are adopting homegrown values. The go-local philosophy is beginning to shape the strategy of huge global corporations as well. General Electric is about as global as you can get, with 60 percent of its revenues coming from Asia, Europe, Latin America, and Africa and 54 percent of its 287,000 employees working outside the United States. In the rush away from manufacturing and toward services in the nineties, the company placed its emphasis on its huge financing operations at GE Capital as profit generator, while outsourcing the making of appliances to Asia and Latin America.

Only after the financial crash of 2007 severely hurt GE Capital—the financial unit was, like so many others, heavily invested in subprime mortgages, real estate, and credit derivatives—did CEO Jeff Immelt change its strategy to focus less on finance and more on making. He changed the portfolio of GE operations, selling off financial businesses, limiting GE Capital's contribution to profits while revamping older factories, buying wind and solar companies, and reemphasizing high-end engineering. GE is spending $1 billion to bring manufacturing back to its plants in Louisville, Kentucky; Bloomington, Indiana; and Decatur, Alabama. Some 1,300 jobs will be added when the refurbishing of the facilities is completed. Water heaters and washing machines, now made in Asia, and refrigerators, currently made in Mexico, will be on assembly in Louisville.

Why the shift to "reshoring"? Lower prices for technology make manufacturing at home in the United States easier. And rising wages in China, Mexico, and other suppliers coupled with falling wages in the United States have made it more advantageous to make things in America. Finally, and perhaps most important, being close to customers who increasingly want to participate in the design of their consumer goods makes manufacturing across oceans and time zones problematic. So Chip Blankenship, chief executive of GE Appliances, went back and reevaluated how and where GE made its consumer products. Soon many that were stamped Made-in-Mexico will be stamped Made-in-the-USA.

GE is also rethinking not just where but *how* it makes things, especially its most advanced high-tech jet engines and MRI machines. Led by Christine Furstoss, Technical Director for Manufacturing and Materials Technologies at the GE Global Research Center, who runs a group of 450 engineers and scientists working on materials, energy storage, and processing technologies, the company found that with more and more parts made of titanium, ceramics, and carbon fibers, there was a huge amount of craftsmanship that went into engines. In fact, in many ways, making these engines had more in common with handmade jewelry than with mass-produced widgets.

GE is embracing 3-D printing technology in the manufacturing of its most profitable products, such as jet engines and ultrasound transducers or probes, often used for medical diagnosis. Typically these devices are made by micro-machining a tiny block of ceramic material—it's hugely expensive because ceramic is harder to drill and cut than metals and a precision shape is necessary to generate sound waves. Printing them is much easier—and potentially cheaper. The process begins with spreading a thin layer of ceramic slurry containing a light-sensitive polymer onto a print table. It's exposed to an ultraviolet light that hardens the polymer. The print table is then lowered by a tiny fraction of a millimeter and the process is repeated—the sensor is built layer by millimeter layer. Once it's done, it's taken out of the small 3-D printer box (about the size of the box your thirty-inch TV came in), cleaned, and baked to sinter the ceramic particles together.

"We are seeing the convergence of the digital and tactile," says Beth Comstock, Senior Vice President of GE. Recently GE "printed" out an entire jet engine, complete with moving parts. It was six inches tall.

We have reached an inflection point, where globalization is peaking and people on a wide scale are beginning to embrace an economy centered on homegrown values. Two of Chrysler's most popular ad campaigns—"Imported from Detroit" and "The Things We Make, Make Us," both created by Wieden and Kennedy—reflect that shift. When Daimler AG, maker of Mercedes Benz, owned Chrysler, its ads focused on German technology and featured the CEO of Daimler, Dieter Zetsche. More recently Chrysler's ads have featured Eminem, the rapper and actor who starred in the acclaimed 2002 movie *8 Mile* set in the gritty neighborhoods of postindustrial Detroit.

In order to participate in the maker economy, we all need a new set of tools. Those of us who know how to work with our hands, or who have embraced digital forms of making—composing well-designed Pinterest pages or Instagram albums with ease—have already mastered many of them. We may not yet see how to use these, and other, tools to transform our lives and careers—but it is possible. And a lot easier than you may think.

LINKING TO A BIGGER PLATFORM

Amy Turn Sharp and her husband, Joe, had been making a living remodeling and selling old houses in Columbus, Ohio, when not just one, but two, toy recalls meant getting rid of toys belonging to her two young sons. "We were so highly annoyed," she told the *Columbus Dispatch*. Rather than accept the possibility that her children's toys might well contain lead paint, Amy and Joe took a different approach. Joe, who grew up in Britain, was trained as a master carpenter. He loved working in wood. Why not combine the artistic and marketing talent she'd developed from the remodeling business with her husband's woodworking expertise to produce a line of natural wood toys for kids?

There is something magical about wooden toys. Boomers and Gen X-ers remember the ones we had as kids: those colorful Tinkertoys and wooden blocks. We treasure the hand-me-down wooden toys of our parents and grandparents and the high prices of vintage and even modern German toys on eBay indicate that we'll pay a great deal for a close approximation.

Amy and Joe began to make teething rings, rattles, and little wooden toys, including a guitar, butterfly, and mustaches, out of natural maple wood grown on a farm in nearby Newark, Ohio, finishing them with organic flaxseed oil. Kids "require so little, but we give them so much crap," says Sharp. "Parents are realizing that less is more, quality is everything, and handmade really means something." She designed some very different, very funky shapes for the toys, combining the traditionalism of old wooden toys with the fun of modern, weird shapes and beautiful packaging. And for every purchase, the couple makes a donation to Tree Pals, which teaches children about the importance of planting trees.

In 2007, the Sharps opened a virtual shop on Etsy, called it Little Alouette, and began selling. It was incredibly easy to do. Their seller's page has a simple design featuring their offerings, and payment can be made through PayPal or credit card. Rattles and teethers went for $12 to $20 and five-block play sets sold for $28. They sold a lot of natural wood toys the first year and raced to meet orders. Their Made-in-the-USA stamp on the toys helped too.

Sharp is an Etsy mom, and there are tens of thousands like her (there are now dozens of websites dedicated to "mompreneurs"). Many begin selling the things because friends or family members ask for them, then go on to expand their businesses to customers all over the world. They set up shop, literally, on Etsy, cut out the middleman, and sell one-to-one to consumers. It's not a bad business, for people who sell on Etsy, for the millions who shop there, or for Etsy itself.

Like so many shops on the site, Etsy was founded by an artisan seeking a direct connection to customers. Rob Kalin dropped out of high school for a time before attending NYU in New York City. His real passion was making furniture, which he did out of his apartment in Brooklyn. But selling furniture proved to be more difficult. Kalin wanted to get directly in front of his customers, but there was no marketplace for his work. So he decided to build one.

In 2005, he got two of his NYU friends to set up servers for a new e-commerce service that would allow producers to share their experiences and challenges with each other. A community of crafters, in other words. This was in the days before Kickstarter, so Kalin asked a furniture customer of his to kick in $50,000 in seed money. Headquarters were set up in the DUMBO neighborhood of Brooklyn.

Six years later, Etsy has evolved into a full-scale platform for "artsy" entrepreneurs. According to the *New York Times*, Etsy has 875,000 sellers and each month the site receives about 40 million unique visitors. In 2011, $525 million in sales occurred on Etsy, with about a third of the buying and selling taking place outside the United States. Etsy now offers tools to let sellers advertise their shops, an iPhone app and Pinterest connection, and versions of the site in French and German. And there is talk of doing an IPO sometime in the near future. Etsy has been able to raise several rounds of financing from venture capitalists; the latest was $40 million from Union Square Ventures, Index Ventures, and Accel Partners in 2012.

The year before, the company got a new CEO, Chad Dickerson, after Kalin was nudged out by the board, which felt the company needed a person with different skills to scale further. Dickerson had been the company's chief technology officer since 2008, when he was brought in to help expand Etsy as a platform by connecting it to Facebook, Pinterest, and other social media sites and to develop its own technology to make it easier for people to sell on it.

As a digital platform that sells tactile goods—wedding dresses, toys, and thousands of other handcrafted objects—Etsy is illustrative of the new kinds of business made possible by the embrace of maker culture and the affordability of easy-to-use digital tools. The costs of opening a "shop" on Etsy are much lower than on Main Street; the site charges sellers 20 cents for each item listed and 3.5 percent of each sale. Etsy's user base consists largely of women, and also provides a support network, including several blogs and forums where sellers swap tips and words of encouragement.

Martha Stewart Media gave an early and big boost to Little Alouette, featuring the company in its online publication in 2010, after it won a Martha Stewart crafts contest.

The toy company then moved deeply into social media, with a Twitter stream, and Facebook, Tumblr, blogspot, and Pinterest pages, all of which direct traffic to the seller's page. Swiss Miss, Cool Mom Picks, and other blogs and sites have featured their wooden toys and celebrated their company. There is a certain irony in this. Even though Sharp doesn't allow her kids, who are homeschooled, to have computers, she knew that utilizing social media would be important to her company's success.

Although the number of people turning to Etsy as a full-time career is unknown, Amy and Joe Sharp are making Little Alouette their full-time occupation. The site does not track how many of its members try to make a living, nor does it disclose the sales figures for individual sellers, said Maria Thomas, former chief executive of Etsy. But over the last year, the number of registered members has grown to 15 million. At the very least, there are many people who are thinking of using Etsy as a platform on which to build businesses.

Scaling was once a hugely expensive part of owning a business, the province of only the largest corporations. With platforms such as Etsy, eBay, and Amazon, scale is now available to all of us. We can simply plug into the "back office" of the PayPal payments system, the community of other craftspeople and entrepreneurs setting up businesses, and, of course, the huge consumer market that the platform provides.

And it doesn't take much effort to link a digital storefront to social media platforms such as Facebook and Tumblr and YouTube in order to personally publicize your business at virtually no cost. Being an entrepreneur, whether you're a craftsperson or not, has never been easier.

Like other start-ups today, Little Alouette is able to be, to borrow a phrase from IDEO's Tim Brown, a "multinational of one" (one family in its case) because its presence on Etsy's global platform parallels the huge operations of multinational corporations, just on a much smaller scale. Amy and Joe can sell the wooden toys they make themselves from sustainable trees harvested nearby not only to their friends and neighbors but to families with little kids in France, Brazil, Canada, and back in China as well. Platforms like Etsy and eBay are helping American entrepreneurs brand "Made in the USA" for export.

CREATING IN THREE DIMENSIONS

Dan Provost, an interaction designer at Frog Design, loved his iPhone camera—easy to use, light, with hundreds of apps that let him play with filters and shape. He is one of the millions who have made the iPhone their main camera. We've all come to treasure the ability to share our pictures with our friends and followers instantaneously. But there was one problem. "Because of its small form factor (and Apple's minimalist design approach)," said Provost when explaining his future new product to an audience of possible users, "it will never contain a threaded nut for attaching a tripod—standard on almost all photo and video cameras."

Rather than lug around a bigger camera for shooting portraits or sunsets, he joined with his friend Tom Gerhardt, an MIT Media Lab–trained hardware and software developer at Potion. Both were designers—Gerhardt worked on interactive installations for retail stores and museums at Potion—but neither had any direct experience bringing a product to market, according to the *New York Times*. They were, however, able to see not only what they loved about the iPhone camera, but what wasn't there. To be a more sophisticated camera, it needed a tripod to hold it steady, so Provost and Gerhardt decided to make a tripod mount that doubled as a stand.

The two began by sketching outlines of their idea by hand. There are dozens of programs, including Google SketchUp, that allow you to create designs that are far more elaborate and complex than ever before. But sketching by hand allowed them to dive right in. "From the beginning, it was clear that simplicity was going to be a key tenet of our design. Not just for philosophical reasons but to keep the design focused and, quite frankly, achievable," said Provost. They quickly came up with a shape—something that attached to the corner of the phone, ran along the long edge, and had a tripod screw.

So what next? The story of how Provost and Gerhardt brought the Glif to market, which Provost details on his blog, *The Russians Used a Pencil*, is just one of countless examples of how any of us, by mastering a few inexpensive, easily available Web tools, can bring our ideas to life.

Just a few years ago, making a new product would have taken one, two, or more years and several millions of dollars to do. For two guys who'd never created a product before, getting in front of angel investors to get seed money would have been extremely difficult and required enormous amounts of networking. If they got the money and designed the product, the prototyping would have been expensive and slow because the machines were expensive and slow. Manufacturing probably would have involved many trips to Asia for the tooling and then the actual production process. How were they going to connect into that outsourcing network of OEMs—original equipment manufacturers—that makes practically everything these days? Orders are backed up for years by the Apples and HPs and Samsungs of the world. How could they squeeze in their relatively tiny order?

Fortunately, Making has become a lot cheaper and simpler.

The price of design software has plummeted, and there are tens of thousands of applications—most of them incredibly inexpensive if not free—that can help us sketch out 2-D designs. But until recently, we couldn't get these prototypes off the screen into 3-D. Now we can. A Brooklyn-based start-up called MakerBot, for example, has on offer a 3-D printer for $1,749 that can build physical objects out of plastics, rubber, wood, aluminum, and titanium. These great printers aren't just for aspiring inventors but for anyone who wants to give his or her possessions a personal touch: If you break your paper towel holder or want a new knob for your toaster, for instance, you can download a file and print a new one on your MakerBot Replicator. But as cool as these printers are, they're hardly necessary. Services like Shapeways and Ponoko will print your prototypes for you. Just send your digital file of the 3-D model of the proposed product and these services will print them out and send them to you within days.

This, in fact, was the strategy of Gerhardt and Provost. The first Glif prototype was created using 3-D modeling software for Mac called Rhinoceros that was free because the program was still in beta. They sent their 3-D file to the Dutch website Shapeways to create an actual three-dimensional prototype.

Anyone can build a rough, cheap prototype out of masking tape, cut foam, and odds and ends, but the detailed prototype really illustrated the proof of concept in a way a crude model couldn't. And because 3-D printing is so cheap and so fast, you can use it not only for prototyping fast but for actual production of hundreds of products. Once it has the digital file, the 3-D printer can print out the same thing over and over again. It's a brilliant new take on the assembly line.

If this still seems daunting, you can just walk over to a growing number of places that will train you to use the tools of prototyping. Today there are "makers' universities" everywhere. Apple stores offer one-on-one training and workshops for a variety of digital tools. And there are online communities that are full of tips on making things, such as Instructables.com, an online open-source portal founded by an MIT Media lab graduate. You can learn how to turn a shotgun shell into a USB stick, how to make a super bouncy ball, or how to use arduinos (small circuit boards that act as modular digital building blocks) to turn just about anything into a robot.

After Provost and Gerhardt got the prototype right, they needed money to start production. Traditionally a team in their position would have tried to set up meetings with investors and deliver their spiel. Instead, they used Kickstarter. "People launching projects can build a deep connection with their audience from the beginning," said Charles Adler, one of the crowdfunding site's founders. "They get the money, sure, but they get early adopters, early fans. They get network connections, e-mail addresses, and consulting advice from the beginning. That makes it all happen."

With this kind of open, entrepreneurial capitalism, performance is key. Even for members of Gen Y who are used to making and posting their own YouTube videos, it can be difficult. The first video asking people to invest in the Glif was a bit "too slick," according to the *Economist*. The second featured the two designers introducing themselves and explaining why they needed help. It had music by the Franks, some speed editing, and a lot of honesty. Provost and Gerhardt asked people to pledge $10,000 to begin making the Glif.

To their amazement, they raised the entire $10,000 in just over an hour after Daring Fireball, a website hosted by tech maven and one of the most powerful Apple commentators, John Gruber, ran the video. And the money kept pouring in. They had expected maybe four or five hundred orders, but in the end, they received $137,417 for five thousand preorders from 5,273 "backers." The top sites for generating business were Kickstarter, their own site, Google, Facebook, Daring Fireball, TUAW (The Unofficial Apple Weblog), Twitter, and the *Economist*. "We were blown away," said Gerhardt. "We turned it into a functioning Web store, with a functioning product."

Though fast, the 3-D printing process wasn't fast enough to keep up with the unexpectedly huge order, so Provost and Gerhardt went searching for a large-scale manufacturer. After talking to six manufacturers, they went with Premier Source, a division of Falcon Plastics, located in Brookings, South Dakota. Why? "From the onset, they just got the project," wrote Provost. "It was also important to us that their facilities were located in the United States," he said. It would be good for marketing and make them heroes to their friends.

It took five months from start to finish, from idea to shipping the product, for two young guys to do something they had never done before: create a successful new product. The Glif sold well enough for the two designers to start up their own firm, Studio Neat. They are designing and making a stream of new products, including the Cosmonaut, a wide-grip stylus for your iPad; and the Frameographer, an app that allows you to make time-lapse and stop-motion HD movies on your iPhone.

Provost and Gerhardt set out to make a new product, but ended up with something bigger: a new company. It's an impressive accomplishment, but one that's becoming more common than ever before thanks to new technology, crowdfunding sites, and online social platforms.

MAKING: CREATING A MORE SATISFYING LIFE AND A STRONGER ECONOMY

Perhaps the biggest barrier to making things is the fact that we don't really have to. You can live a comfortable life without ever lifting a hammer or directing a video or even making dinner from scratch. But all around us are hints that a life without making simply might not be as satisfying as one in which we do not just consume, but also create. It's there in the smell of the freshly baked pie at the farmers' market, in the retro-style surfboard your friend designed, in the beautiful wooden door your neighbor rescued from a local recycled lumberyard and refinished.

It's certainly there in Matthew Crawford's motorcycle shop. As he writes in his acclaimed book *Shop Class as Soul Craft*, "Seeing a motorcycle about to leave my shop under its own power, several days after arriving in the back of a pickup truck, I suddenly don't feel tired, even though I've been standing on a concrete floor all day. . . . The wad of cash in my pants feels different than the checks I cashed in my previous job. It may be that I am just not well suited to office work. But in this respect I doubt there is anything unusual about me."

The first step in becoming a maker just might be imagining the satisfaction that can come with creating something. You may feel you're too busy to do anything but crash after a long day of work, but ask people who've embarked on a project from start to finish and they'll tell you that the act of making something and sharing it with family, friends, or customers leaves them much more invigorated than sitting in front of the television or ordering in takeout.

The next step is realizing that making things—all kinds of things—is a lot easier than you think. Today we all have access to a nearby "Make U," a place where we can go to learn how to use technologies both new and old. Yes, there are a handful of specialized universities for the few fortunate among us who sought out a design education early on. Places like the Rhode Island School of Design, the MIT Media Lab, and the Royal College of Art in London have the latest equipment and experienced teachers—but they're hardly the only places you can get a stellar apprenticeship in the world of making.

Where previously, high costs of a specialized education would have prevented experimenting with different media and making technologies, today, for the price of an iPhone, you can go to an Apple store and learn to use incredibly advanced digital tools. A new generation of "shop classes" are being offered in major cities by companies like TechShop that teach the tools of digital fabrication.

If your tastes lean more toward traditional crafting, the opportunities there, too, are growing. For those interested in the locavore food movement, there has been a surge in classes throughout the United States not only on how to cook nearly any cuisine, but also on how to butcher your own pig, hunt for deer, or forage for your own ingredients. If you have a number of interests and aren't certain which one to focus on, you can try your hand at a number of skills or trades before investing a ton of time and energy into any of them.

For those of you who've already begun creating and want to take what you've made to market, or to a larger one, you can expand your repertoire to include presentational skills that can bring your idea to a wider audience. For years, business gurus have spoken about the need for a great "elevator pitch" to sell your idea to higher-ups or investors. But in an era of social media, the "pitch" is changing as the stage shifts from elevators and boardrooms to PC screens and mobile phones. The thirty-second pitch is being replaced by the one-minute YouTube performance. A strategy for marketing your pitch to hundreds of thousands if not millions is the key to selling your ideas, getting funding to make them happen, and expanding your networks.

Even if you own or work at a bricks and mortar store, you need to think about presentation. The cost of high-quality video equipment and user-friendly editing programs has come down so significantly that it doesn't take a lot of time or money to put together a quality pitch or presentation of your product or idea. Framing strategies can help you make presentations that tell the right story and speak directly to the audience you're trying to reach. And thinking about playing and gaming strategies, too, can be key: How can you make potential audience members and customers part of the journey?

You can also bring the things you've made to larger audiences by exploring established platforms like Etsy, eBay, or Amazon. If you've sold something, either as an individual or through your business, you know how important platforms can be in getting what you make into the hands of consumers. New platforms make setting up a storefront a snap.

While there's never been a better time to start exploring the rise of maker culture, it might take a little time for you to discover exactly how to make best use of your skills, experiences, and areas of interest. And of course many skills take time to learn. In the meantime, you can keep your enthusiasm going by supporting homegrown businesses or checking out nearby markets or Maker Faires. Talk to farmers, furniture makers, or other artisans in your community. Ask them questions about the benefits and the challenges of their work. Get educated and inspired. You may start out as a consumer only to notice an obvious gap in the market. Sometimes creative inspiration comes from looking at existing products and thinking, "I could do something like that . . ."

With companies as global as GE and Caterpillar moving production back to the United States, it's no longer possible for global corporations to ignore a growing demand and opportunity for making physical goods in the United States. I do not underestimate the amount of time it takes to reassess global outsourcing strategies nor the complexities of embracing these shifts, but as outsourcing grows more expensive, machines that make things get cheaper, and Americans are beginning to demand that more stuff be made at home, exploring more local, sustainable production is an obvious part of any multinational strategy.

Of course, it's important to keep in mind that a strategy that relies on outsourcing carries with it an even greater potential danger than the growing costs of foreign labor and transportation: the risk of losing a generation of talented young people who want to design great *stuff*, not just strategy. If you don't give your bright young engineers and designers the chance to be more involved in the entire design process of some life-changing new product, some start-up will.

Companies have been outsourcing for so long that many managers assume it's the best way to make things. A whole generation has grown up with outsourcing as its primary manufacturing model. But now that labor and energy costs in China are rising and technology and manufacturing costs in the United States are decreasing, it's a perfect time to ask some simple questions: If we wanted to make things in the United States again, how would we do it? Are there still suppliers in this country that we can count on? My guess is that there will be a lot of surprises, that people who ask these questions will discover that a significant amount of knowledge has survived the wide-scale shift to global manufacturing strategies. After all, the whole supplier network ecosystem of the Detroit auto industry survived, which is why auto manufacturing in the United States is booming.

But perhaps the most important questions to ask are: Is there advanced manufacturing technology that didn't exist thirty years ago that could enable us to make things better today? Can we make things in an entirely different way than we did before? There are new opportunities in many "old" industries like furniture, carpeting, and apparel—industries that shrank in the United States as they exploded elsewhere around the world. Think of using an iPad app to design your own carpet, having it made on a high-tech loom, and delivered the next week to your door. It's already happening with T-shirts and running shoes. In what other industries can we move from mass manufacturing to mass customization?

Pivoting

As a member of the seminal gangsta rap group N.W.A., a pioneer of the sound known as "g-funk"—the West Coast rap style heavily influenced by George Clinton and Isaac Hayes—and a producer who has worked with or helped launch the careers of a number of the biggest rappers—Eminem and 50 Cent, to name just two—Dr. Dre is considered one of the most influential recording artists and producers in hip-hop.

He's also known to be a perfectionist in the studio. But you might not know the full extent of his hard work if you listen to his songs on standard headphones. "People aren't hearing all the music," Dre was quoted as saying. "Artists and producers work hard in the studio perfecting their sound. But . . . most headphones can't handle the bass, the detail, the dynamics. Bottom line, the music doesn't move you."

In his three decades making music, Dre has created an empire around his brand—as of 2012, he was ranked third on the *Forbes* list of the richest hip-hop figures. But what good was his brand if the majority of people listening weren't really hearing what he was creating?

Dre didn't just want to make a better pair of headphones, he wanted to connect his fans to the music as it was created in his studio and played in the clubs. He wanted them to hear his sound profile, heavy at the bottom but with clarity and good range. He couldn't make it happen alone. He needed to partner with some-one who had a track record of bringing ideas to life.

THE MAN DRE TEAMED UP with was product designer Bob Brunner, though before I discuss his work with Dre, it makes sense to share a bit about how Brunner came to develop that skill.

Brunner is famous in design circles for many things—among them, hiring Jonathan Ive while director of industrial design at Apple in the nineties. By 1996, he was partner at lead graphic de-sign consultancy Pentagram in San Francisco, heading up a team offering product, identity, and design services to clients.

While at Pentagram, Brunner was approached by the Discovery Channel to participate in a show about the design of a real product. The cable company wanted to take its audience inside the process, to show how a product is actually created and made. Brunner was game, but client privacy clauses prevented this kind of inside look at the projects Pentagram was working on. Brunner came up with a solu-tion: Why didn't the firm just make something up? Brunner and his team rose to the challenge and invited the Discovery crew in to spend time with them as they developed something entirely new.

But what should they design? What would the Discovery Channel audience relate to? Well, the team reasoned, they did live on the West Coast, where people loved to cook outside, so why not do what no one had ever done before: redesign the old, tried and true outdoor grill?

The team went to work, observing how people used their barbecues at parks in the Bay Area and in their homes. "Barbecu-ing is a social event, and the barbecue a social hub," said Brunner. "There's a fire, there's food, and people like to crowd around." And yet, the team realized, most barbecues are designed so that the chef has his back to his guests (yes, they learned, men still did most of the grilling).

The team decided to make a new kind of grill, one built with great materials but, most important, *round*—a fire pit that people would love to gather around. With the Discovery Channel cameras in mind, they set out to create something that would stand apart from the typical drab forest green grills. Something beautiful. So beautiful that both the Pentagram and the cable television people said they wanted one when the grills were done.

Normally Brunner's job would have ended there—but it didn't. Like many top designers, Brunner had long felt that those in his field didn't really participate in the full value of what they created. Designers got paid for the services, sure, but then another company typically would take over, manufacture the product, and sell it—often for big profits. So Brunner and his team decided to go further with the design of the grill, which he owned, by starting a new company, Ammunition.

Thanks to Discovery, Brunner and his team designed a series of grills, set up manufacturing in China, came up with a great brand name—Fuego—and launched a new company. Priced from $350 to $700 and sold at Williams Sonoma and other high-end stores, the grills were a success.

Brunner now has two partners, Brett Wickens and Matt Rolandson, as well as forty-five people working at Ammunition. He still does regular design service work for some clients but prefers to partner up or start new companies. "It's much easier to get things made today given the lower cost of outsourcing, things being made in China and around the world, and lower costs of technology. The channels are already there and there's no reason you can't do it."

IN A VERY ORGANIC WAY, Brunner made the pivot from creating new products to participating in the creation of companies. But he's hardly the only one. So pervasive is the change in the nature of creative work that designers everywhere are becoming entrepreneurs. Not only have a large number of today's leading startups been founded by people with design backgrounds, but walk into almost any classroom in the top design schools in America and Europe and you will find the majority of the students planning to start their own companies upon graduating. Many major design schools now offer courses on entrepreneurialism. Stanford, of course, has been doing this for decades, but an entrepreneurial spirit is spreading throughout the academic design society.

Venture capitalists and other investors have taken note. In 2011, the Designer Fund was set up by a group of successful designers to help would-be designer-entrepreneurs turn their ideas into businesses. The real draw of participating isn't the financing offered, but the network: The members of the Designer Fund match young would-be entrepreneurs with mentors and the whole constellation of start-up capitalists—angels, seed funds, and later-stage venture capitalists like Andreessen Horowitz and Khosla Ventures.

This pivot from creativity to business creation is one of the strongest economic forces of our time, but it isn't just happening with the fields of design. There is an explosion of entrepreneurialism in America today, especially within that Generation called "Y." Many young people are making the leap from curating their digital spaces and creating physical products to launching businesses and social organizations. About 180,000 master's degrees in the field of business are given every year and, increasingly, students are focusing on starting their own companies rather than managing existing ones. Courses in entrepreneurialism are among the most popular in B-schools at Columbia, Harvard, Wharton, Chicago, the University of Toronto, and others trying to keep up with Stanford's wild success. Harvard Business School, a long-time training ground for the corporate elite and consultancies, recently opened the $25 million "i-lab" or Arthur Rock Center for Entrepreneurship (named after an HBS alum who invested in Intel and Apple). Where in previous decades graduate-level business programs focused on how to use capital efficiently, more and more courses now focus on how to harness creativity.

The widespread pivoting from product concept to business creation is beginning to revive and remake cities. Richard Florida has long discussed the role of the "creative class"—the 40 million or so working in the fields of design and architecture, art, media, entertainment, science and technology, education, and health care—in driving the innovation and economic growth in cities. But even Florida might be surprised at how fast creatives are transforming such giant cities as New York, Berlin, and, perhaps, even staid Singapore.

A 2012 study for the Center for an Urban Future called New Tech City concluded that in the past five years, New York has become the country's second leading hub for tech start-ups after Silicon Valley. Since 2007, local venture capital firms such as Union Square Ventures and IA Ventures, as well as other investors, had funded nearly five hundred digital start-ups, the Gilt Groupe and Tumblr among them. Tumblr CEO David Karp describes the New York start-up scene "as very design-centric and very media-centric." There were around a dozen tech incubators in New York City in 2012, where in 2009 there had been only a few. Go to a meeting of NY Creative Interns, a network set up to promote entrepreneurship among creatives in New York City by Emily Miethner, and you'll see up to two hundred recent graduates scurrying to make connections to help them launch their ideas.

Even New York Mayor Michael Bloomberg, who began his career at Salomon Brothers and whose financial data company provides sophisticated analytics on "Bloomberg Terminals" to more than 300,000 professionals around the world, has become an advocate of start-ups. In 2011, he set up a contest that resulted in a partnership between Cornell University and Israel's Technion (Israel Institute of Technology) to build a new engineering campus. The hope is that it could rival Stanford, with its success at spinning off high-tech start-ups. For some time, policy makers like Bloomberg have viewed the nation's economic future through a finance-centric lens; dot-coms and start-ups were certainly fascinating but considered by many to be peripheral to real economic growth. The core courses at nearly all business schools focused on managing operations and marketing, not starting new companies. Students were taught how to use capital to grow, not create new things entirely. Now the direct connection between the creative nature of start-ups and capitalism is beginning to be made central again.

The shift in the focus of economic thinking back to entrepreneurial creativity presents the questions: How do you transform an idea into a thriving business? What can you learn from designer/entrepreneurs who've made that pivot?

The word "pivot" is often used in Silicon Valley to describe the move from one idea to another in the early stages of a start-up. Often founders start out with one idea, only to "pivot" to a second or third before hitting on the product that brings them great success. Instagram, for example, started out as Burbn, a location check-in app for smartphones similar to Foursquare. Burbn didn't succeed, but the photo posting function that founder Kevin Systrom built into the app got a lot of traffic, so Systrom changed strategies and built Instagram.

I'm using "pivot" in a larger sense, to describe the movement from inspiration to production, often production on a wide scale. Pivoting is the scaling of creativity that's essential to creating new products, new models for business and nonprofit organizations, and even entire industries. How and when to Pivot are the key strategic questions every entrepreneur, leader, or aspiring creator should be asking today.

On a deeper level, Pivoting involves taking the intangibles that money can't buy—our dreams, our desires—and turning them into the things that it can. Artists have always known this—painters like Edvard Munch, musicians like Patti Smith, writers like Kurt Vonnegut have been willing to transform their deepest hopes and fears into something that speaks to us all. So, too, have creative people in every industry. That's what creativity can do, create gold from straw, art from angst, and yes, household products from wishes for a better life; in that way, it's a kind of alchemy. But no economic model analyzing innovation has ever factored in these intangibles.

In fact, many assume that "what money can't buy" and "what money can buy" are mutually exclusive categories. In his book *What Money Can't Buy*, based on his popular Harvard class, Michael Sandel attributes this to the conflict between social norms and market norms—and our discomfort at mixing them. Yet asking yourself "what money can't buy," and using your answers as a source of inspiration for creating new products and services, has always been intrinsic to capitalism and necessary for economic growth.

Pivoting from creativity to creation bridges the gap between social and market norms, and so it requires skills that we may not ordinarily associate with economics or business. At a time when people are no longer as willing to believe an advertiser's passionate pitch about a new widget, today's creative thinkers are embedding deep meaning into the products they create, imbuing their products with a kind of aura typically found in great works of art. And leaders are adopting a deeper kind of charisma—one that has less to do with flash and more to do with deep connection to their followers.

Of course, people have been pivoting from concept to creation since the first goods were ever sold, and so there is wisdom to be gleaned from looking back at some of the great innovation stories that predate our current era of entrepreneurial capitalism. Whether they sought out inspiration from a muse, support from a general manager, or the embrace of an audience, creative innovators have rarely worked alone.

AURA: EMBEDDING MEANING IN YOUR OFFERINGS

Music moves people. It lifts them out of their everyday moments and takes them to another place. Sales of the headphones that Dre created with his longtime partner Jimmy Iovine, founder of Interscope-Geffen-A&M Records, and Bob Brunner are soaring because they are giving fans of hip-hop the gift of music as it was meant to be heard. The name of the audio company came directly from the source of the sound. "Dre came up with the name Beats," according to Brunner, because "he said 'people come to me for my beat.'"

Dre and Iovine shaped the sound profile they wanted to come out of the headphones (and later earphones, speakers, and boom boxes) and worked with the manufacturing and distribution teams at Monster Cable to develop them. The headphones—there are models for consumers as well as for professional DJs—generate deep, rich bass. And the design, too, was key: the Beats are decidedly rapper nasty, in a wide range of color: black, moody red, hot pink, white. They're made from soft rubber, sculpted metal bezels, and soft-touch plastics that feel good in the hand. Brunner also helped design rich, intricate packaging for the Beats—a box has multiple compartments and folding doors that open easily, but slowly. Opening the box becomes a kind of ceremony culminating in the gift of authentic music.

Dr. Dre has amplified the Beats community by getting big-name athletes and musicians to wear his headphones; in August 2012, the NPD Group reported that the brand accounted for around 50 percent of the premium headphone market. A Beats store just opened in Manhattan's SoHo neighborhood and, like Apple stores, it will showcase just a few products. The company is now looking to expand its success beyond headphones; in 2012, Beats Electronics bought Berkeley-based music subscription services MOG, giving Dre a way to deliver music directly to his listeners; that year Beats also dropped its association with Monster, and Taiwan-based HTC became a big investor. If current projections continue, Brunner believes that Beats will be a $1 billion company soon.

It's perhaps no secret why the Beats have taken off: a hip-hop legend teamed up with a renowned designer-creator to build a unique set of headphones that delivered a hip-hop sound. But far more important, Brunner and Dre built a special bridge between Dre's music and his audience. After using the Beats headphones and hearing, for perhaps the first time, the sound that artists like Dr. Dre set out to create, people sense they are connecting to something that transcends the everyday.

Which, of course, is a lot to say about headphones. Yet by creating a product that takes people closer to the moment of creation in the studio, the team that worked on Beats is giving people the opportunity to have "secular epiphanies."

In his 1933 essay on Japanese aesthetics, *In Praise of Shadows*, novelist Jun'ichiro Tanizaki describes a similar transcendent experience—albeit one triggered not by new technology but by something more traditional: rice. "A glistening black lacquer rice cask set off in a dark corner is both beautiful to behold and a powerful stimulus to the appetite. Then the lid is briskly lifted, and this pure white freshly boiled food, heaped in its black container, each and every grain gleaming like a pearl, sends forth billows of warm steam—here is a sight no Japanese can fail to be moved by. Our cooking depends on shadows and is inseparable from darkness."

Tanizaki doesn't use the term "aura," but it's the best word I've come across to define that powerful engagement we have with certain objects and products. Walter Benjamin argued in his 1936 essay "The Work of Art in the Age of Mechanical Reproduction" that unique works of art have an ineffable quality to them, an "aura." They beckon us. Paintings, sculpture, live performances all share a certain aesthetic quality that we can experience face-to-face in museums or theaters but that cannot be reproduced when a work is copied. Cinema and other reproducible art forms such as photography, he argued, lack aura because the process of "mechanical reproduction" disconnects the viewer from the original work of art. But the technology of our time—their improved features and lowered costs, their ability to make us all creators and not just passive users—can, in fact, connect people in ways that the films or photographs of seven decades ago could not.

As with many of the Creative Intelligence competencies, the road leads back to Apple. Consumers call Apple products "cool" and "easy to use," and more sophisticated business analysts applaud Apple's "ecosystem" of integrated software and hardware. But none of those qualities alone explains why we feel the way we do about Apple products; it's impossible to discuss Apple products without mentioning how they feel in the hand, look to the eye, and connect to our deep emotions. The story of how Apple began creating beautiful, easy-to-use products should be required reading for anyone interested in creating something that's not just useful but meaningful.

WHEN STEVE JOBS RETURNED TO Apple in 1997, after twelve years in exile, he bet the company and his future on a radical new idea: an easy-to-use, stand-alone PC that looked unlike any other computer before it—translucent, colorful, fun. As much as we hail Apple for its cool design these days, so many other companies have stepped up with well-designed products of their own that it's easy to forget how very *not fun* computers used to be. They were hard to use and ugly to boot, all of them some variation on the color putty.

But no one in the late nineties knew how to make a translucent casing for a computer, especially one with colors. And certainly no one knew whether you could make these things for a profit.

Designers Jonathan Ive and Danny Coster looked beyond the world of computers for inspiration until one of them, it's not known who, came up with the answer: jelly beans. They're colorful, translucent, and manufactured in large scale—not to mention they're one of those rare things that can bring a smile to just about anyone's face. After making this observation, Ive's team of designers and engineers spent time in a candy factory, analyzing how jelly beans are made, examining how bright colors are applied to shiny surfaces, and how hard plastic jelly beans are extruded by the tens of millions from specially designed machines.

Next the team traveled to Japan to meet with the best injection-molding tool makers to configure a new way to inject molten plastic and metal through tiny feed lines, getting the right number of holes so that it would cool to a perfect surface in seconds. Ive then spent yet more months in Asia with manufacturers, pushing factory owners used to "faster and cheaper" to focus on quality. In the end, Ive's team devised sophisticated processes for manufacturing millions of jelly bean–like iMacs. (And, along the way, in a little-told tale, Ive and Apple pretty much transformed China from a nation known for cheap and shabby exports to one able to make computers, TVs, and, later, iPhones and iPads of the highest quality.)

iMacs were an amazing surprise. These brightly colored, round machines were warm and friendly, the opposite of the lifeless, putty-colored boxes on office desks. But Ive's team didn't stop at designing the iMac. They also designed a beautiful "reveal." The process of taking the computer out of the box was designed to be a delight. It was simple and easy—a ritual that people wanted to participate in. The packing materials themselves had a Bauhaus feel—clean, simple lines with materials that felt good in your hands.

The iMac's launch in 1998 marked Apple's return to profitability after years of being in the red and helped make one of the men responsible for the amazing look and feel of Apple products, Jonathan Ive, a hero in the design world and beyond.

In a 2006 interview with Peter Burrows, who covered Apple at the time for *BusinessWeek*, Ive shed some light on Apple's process. "We don't make very much stuff," he said. "That's a very important part of our approach to what we do, which is to not do a lot of unnecessary stuff but just to focus and really try very sincerely to care so much about the few things that we do. We just make lots and lots of prototypes. Then we spend a lot of time at the manufacturing sites. We'll be there right to the end when we're in production."

As I write in late October 2012, Apple is the world's largest company by market value, with stocks valued at around $650 per share for a market value of over $600 billion. But financial numbers alone don't tell us anything about the "aura" of Apple, about why people so identify with its products that Jobs's death seemed almost to be a national day of mourning, as numerous memorials appeared outside Apple stores the morning after his death was announced.

As Walter Isaacson's biography reveals, Jobs was not a nice guy or an easy person to get along with, especially if you were close to him. Yet he wanted to instill a level of aesthetic perfection in Apple products so that they would remain distinctive and never fall into the rut of "mechanical reproduction" or pure salesmanship (which is where he saw Microsoft going under Steve Ballmer's leadership). He saw himself as a designer of things that people didn't even know they wanted until he created them.

But there's more to Apple products than their beauty. Their common aesthetic suggests that they are connected to one another, a "family" of products. The iTunes app acts as both a commercial and a community hub, allowing music and other entertainment to be purchased with ease, and linking all the products to the ever-evolving "cloud" that represents designers' dreams of connectivity. Each app allows consumers to "personalize" their access to Apple and interaction with other users. And the recent addition of Siri adds a new sense (voice) to touch and vision, re-creating a part of that face-to-face dimension so important in Benjamin's account of ritual aura. What Apple has done (and what Google and Facebook are trying to do) is involve the individual customer and wider community in an open-ended process of creating.

But, of course, a company cannot live on what it's already created alone. It has to continually maintain the aura of its products to keep the company going and support its creativity. We have recently seen a spate of what seemed to be innovative companies fail in maintaining that aura in their products, and there is no guarantee that Apple will succeed despite Jobs's assurances that his successor, Tim Cook, understands the need to have goals that transcend simply making a profit. Certainly Apple's aura has been compromised because of labor conditions at Foxconn, the Chinese manufacturer of iPads and iPhones, as well as the less than perfect launch of Apple Maps.

The power of aura cannot be underestimated. I remember seeing a Ferrari designed by Italian firm Pininfarina at a conference in the mid-nineties that mesmerized me. After the speeches, I broke away from my colleagues and draped myself over the hood of the Ferrari and announced that I had "product lust." I was swept away by the Ferrari and its promise of beauty, power, celebrity, speed, excitement, whatever. That level of engagement is the reason why Apple fan boys and girls wait in line for hours for the newest iPhone, and why so many aging boomers won't ride anything but Harleys. When making the pivot from creativity to creation, thinking about how people connect with things is crucial to the success of the effort.

But what happens next? What if you have the idea but don't know whether, or how, you should pivot and turn it into something real? Whether you are working in a big company or alone in your studio or apartment writing or painting, you need someone trained to spot creative ideas and help turn them into creations. You need a wanderer.

FIND YOUR WANDERER

In their early days, many of the big corporations we now consider traditional, risk-averse, and efficiency-focused were every bit as creative as today's start-ups. They were younger, more in touch with their roots, and connected to the culture of their founders. They could jump on new creative ideas and pivot them into wildly successful products and brands. And many are trying to achieve this same creative spirit today—Boeing has begun using new composite materials for its 787 Dreamliner, IBM has moved into crowdsourcing and wiring cities for innovation, and Corning is developing new glass for iPhone screens. But few companies have enjoyed the success that Hewlett-Packard had in its heyday.

From its founding in 1939 in a Silicon Valley garage, HP had the kind of culture that today's start-up entrepreneurs would love to emulate. Managers gave their employees, especially the engineers in HP labs, the freedom to play, to mine knowledge from sources that interested them, and to frame ideas however they wanted. Just as important, HP provided a network of wandering general managers who moved from lab to lab, screening inventions and deciding where to invest. These wanderers helped lift new ideas off the drawing board and transform them into reality.

That shouldn't be a surprise to anyone who knows the history of Silicon Valley. The open, collaborative culture we associate with companies from Google to Facebook was modeled in large part on HP—or, rather, on the creative culture and organization built by HP's founders William "Bill" Hewlett and David Packard. The culture of freedom and trust remained intact even four decades after they began the company in which some of the greatest technological advancements of the last century were created, including one that made printing available to the average consumer at a low cost—and without inky fingers.

The story of the HP's ThinkJet is a prime example of how people from different disciplines can work together to create something few imagined possible. The people who helped make the inkjet printer a reality had, among them, advanced degrees in electrical engineering and high-temperature physics, and experience with "integrated circuitry" and "photolithography"—but it was a college drop-out named John Vaught who came up with the idea for the breakthrough technology and then pursued it passionately until it became a reality.

Despite his lack of traditional engineering education, John Vaught loved working at HP. For Vaught, a self-taught engineering associate and technician, "HP Labs was a wonderful place. I had to work in a single field for only two or three years and then, like magic, it was a whole new field; a paradise for creativity." An ideal environment for a man who said of himself: "I bore easily."

According to Lee Fleming, Vaught's partner Dave Donald had a more conventional engineering background, his attention to details a contrast to Vaught's tendency to go "very far, very fast." Like so many partnerships, they complemented each other.

In 1978, Vaught and Donald has just finished adopting Canon 2680 printing technology for HP in the company's big Boise lab and were returning to Palo Alto to begin work on an electrostatic gravure printer for the commercial publishing world. But Vaught kept thinking that the real prize would be a low-cost printer that could deliver much higher quality than the conventional dot matrix.

From the beginning of what we now call the Computer Revolution in the 1960s through most of the 1980s, offices used dot-matrix printers. Remember, this was before computers became "personal" and migrated into our home offices. The technology of dot matrix was pretty simple—pound it out. Dot-matrix printers were "impact printers"—mechanical machines that worked like typewriters: Pins struck an ink ribbon to form characters on a page. But the simplicity of the mechanism was reflected in the results: Printing was slow and loud, and the highest resolution was 100 DPI. Most researchers at HP and elsewhere thought the solution was to put more dots into the existing dot-matrix printer, but Vaught wondered if there was another solution entirely.

HP didn't start out in the consumer printer business—and it certainly wasn't in the low-cost printer business. In fact, the joke going around at the time was that "HP" stood for "high price."

For most of its early history, the company made scientific instruments like oscilloscopes, computers, and calculators. As HP's Frank Cloutier later reflected, in the late seventies, around the time Vaught began thinking of a dot-matrix alternative, "We weren't the largest printer company on the planet. . . . We had no business going into that as an enterprise."

And yet on Christmas Eve in 1978, when, according to HP tradition, families of the engineers and researchers came to the labs and offices to celebrate the holiday, Vaught, Donald, and the team of engineers began to throw around ideas about their ideal printer. Their goals included color imaging and a printing speed of a page a second.

The meeting eventually ended and the families went home for the holiday, but it was after the break that Vaught came up with the idea that has become HP legend: According to Fleming, as Vaught caught sight of a coffee percolator he kept on his desk, he watched the coffee heating up, exploding through small holes in controlled bursts.

It took a while for Vaught to work through the implications of the coffee percolator. It wasn't just a "eureka" moment for Vaught. "Inventors just don't go home and see it at that moment in time," he recalled later. "When it comes to the moment of truth, you think about a lot of things."

The percolator gave him the image and the idea of heating a liquid to the point of explosion. With the percolator, "if you think about it, if you left the top off, it went poof, poof, poof, and blew gobs of coffee all over the place," he said. (Because of this explosive process, according to Thomas Kraemer, the inkjet project would later be named "St. Helens" after the nearby volcano.) But the explosions would have to be directed, controlled. The printer would need nozzles to work.

And so the real work—the pivot into creation—began. Of

course, to Vaught and Donald, it didn't feel like work. I've known a number of scientists and researchers, and I've always been struck by their glee at working in labs. A lab with the right team of people who trust each other to play at experimenting and discovering new things is the perfect example of a magic circle. The way that Vaught and Donald described their time was no different. "They had tremendous fun," wrote Alan G. Robinson and Sam Stern in their book, *Corporate Creativity*. "Like children at play, they were full of enthusiasm, trying first one thing and then another." They used resisters to heat up and vaporize ink, causing it to shoot out microdroplets onto paper. The ink passed through tiny holes in a nozzle, which controlled the flow. In three months' time, they had built a working prototype.

The process promised to be fast, clean, and simple enough to mass-produce at a low cost. And it was also patentable, which was always important with Japan competing fiercely in printers. Vaught didn't know it at the time, but across the Pacific, Ichiro Endo, an engineer at Canon, a leader in printers, was also working on a process to heat up ink and shoot it onto paper. An accident had led Endo to direct his research along the same lines as Vaught two years earlier—while someone was refilling an old inkjet, the ink syringe came into contact with a soldering iron, the heat of which caused ink to splash out. Years later, of course, HP and Canon became close partners.

John Vaught may have bored easily, but he "carried the ball in selling the innovation," Donald told Lee Fleming. According to the *Economist*, "Vaught doggedly pursued his interest in the inkjet printer. He demonstrated his work to anybody who took an interest."

Despite his enthusiasm, there was a problem with Vaught's process. He couldn't really explain *why* it worked. He couldn't articulate the physics of the process, the science part. "Because its inner workings were not understood, even a number of people who actually saw the device operating told him the approach could not work," wrote Robinson and Stern. Later, when university scientists had a look, they described it as a "phreatic reaction." But that was a year away.

Without an explanation, Vaught couldn't get the attention of his general manager—the person who could finance his concept and make it a reality. Eventually, Vaught's manager reassigned him, leading to what he later described as "the worst period of his life."

Meanwhile, Frank Cloutier, a general manager who worked at the HP Corvallis campus in Oregon, was charged with finding something new for HP to make. In 1979, there was extra capacity in HP manufacturing and a need to fill it. Because Corvallis had also been involved in developing printers for HP's portable calculators, Cloutier began to search for something new in the printing field.

Cloutier's role was to support innovation by wandering around. HP at that time had a lot of wanderers. When Hewlett and Packard set up the company, they routinely walked around the labs, checking out what was happening, connecting people, looking for ideas that might be transformed into new products. This later become formally known in management consulting circles as MBWA—management by wandering around. It was a way to cut through the formal organizational hierarchy of big companies that divorced managers from their most productive employees. But at HP, it was simply the way the founders had always done things. It became known as the "HP Way."

The key was to keep the labs autonomous and the engineers free to experiment as they saw fit. But the essential link in pivoting from concept to product was the general manager. The GM, whose role was not unlike a modern VC, was very close to the engineers and had a lot of leeway in allocating resources to develop new products. Most GMs were engineers themselves and they knew one another. Throughout HP, there was friendly competition between engineers as they developed new technologies.

After traveling to HP's senior research lab in Palo Alto, Cloutier found what he was looking for in Vaught's percolator-inspired thermal printer technology. And he knew members of the team he'd just been working with were well-qualified to take on the task of bringing the idea to market. But first they had to figure out the physics of Vaught's thermal printing. Just how exactly did the vapor explosion work that sent the hot ink onto the page?

Thanks to help from colleagues like Larry LaBarre, a longtime HPer who had been drumming up support for the idea, and a researcher named John Meyer who helped Vaught make a pitch for resources, Vaught received $250,000 to work on the project, as well as a recommendation to bring in high-temperature physicists from the California Institute of Technology, who were instrumental in getting the technology right.

Typically, Cloutier reflected, a project like this would not be considered until "technological performance and manufacturing promise are well-established." It became clear that in the case of this new project, an entirely new approach to managing and scheduling would be necessary in order to meet their timing goals: "One was unrivaled communications between the product and technology groups," wrote Cloutier. "The other was an extremely strong commitment to a true team approach, where the whole was clearly equal to more than the sum of the parts. The phrase 'that's not my job' was sufficient grounds for termination from the project."

In a 2004 speech at MIT, Cloutier recalled a meeting that took place as they were developing the new printer. After a number of people threw out goals for the product, one of his team members put a recent *National Geographic* magazine with a beautiful toucan on the cover and said, "We want to do *that*," Cloutier said.

Cloutier agreed. Though the toucan picture inspired different goals among different members of the team—to some, it represented scalable fonts; to others, full color—Cloutier said the image became a "galvanizing vision" for the team. "As you think about visions, you can . . . look at the past and you just extend that linearly into the future," said Cloutier in his MIT speech. "But rather than that, when we said we wanted to do the cover of *National Geographic*, it set the bar at a very different place. . . . And it meant we had to do things we didn't initially recognize we had to do."

It took the work of hundreds of people to bring the printer to market. And it took time—it wasn't until 1984 that the company's first inkjet printer, the ThinkJet, was manufactured, ending the reign of the noisy dot-matrix printer, just as the pocket calculator, the HP 35, eliminated the need for slide rules. That year, HP also released its first laser printer, the LaserJet. Only in 1988 did HP get the price down below $1,000, and its first color inkjet was not released until 1989, but the printer went on to become one of HP's most profitable products. About 300 million have been shipped since 1988. In 2011, HP's Imaging and Printing Group had revenues of $26 billion, much of which came from sales of the ThinkJet.

A month before the launch of the revolutionary new product, John Vaught resigned, freeing himself to work on whatever he wanted.

SCRATCH BENEATH THE SURFACE OF a creative organization, business or nonprofit, and you'll find a wanderer. Steve Jobs was, of course, a lifetime wanderer: He wandered into the calligraphy classroom at Reed College after dropping out to pursue what interested him. He wandered into Xerox PARC in 1979 to see its work on graphical user interfaces and what we would come to call the "mouse." He wandered into the animation studios of Pixar when he was in exile from Apple. And when he returned, he wandered into the studio of his chief designer, Jonathan Ive, nearly every day, to see what new ideas were percolating.

The reason for linking up to a wanderer is simple. Pivoting from creativity to creation requires scale. That scale comes in two forms—capital and markets. Creators by themselves usually don't have access to capital or markets. But wanderers do. It's their job to provide scale, to offer financing and connections to an audience.

These key people, of course, don't call themselves wanderers, but they are everywhere. Talent scouts, coaches, lab chiefs, VCs, curators, agents: people who make their living bringing ideas to life. Without Peggy Guggenheim, who in 1945 financed Jackson Pollock's move to the Springs in Long Island where his Drip paintings were created, the artist may never have developed his unique style of abstract painting. Guggenheim also introduced Pollock's work to the art world in her gallery. Of course, Guggenheim played a pivotal role in the lives of many artists. "It seemed to me she was like an open sesame. She was full of plans," John Cage was quoted as saying. "Peggy had the keys to the whole art world." More recently, Stanford University computer science professors Daphne Koller and Andrew Ng may not have gotten their online education start-up, Coursera, off the ground were it not for $16 million from John Doerr of Kleiner Perkins, one of two VCs who invested in the platform that delivers courses in a variety of subjects in short video episodes.

If finding the right professional wanderer proves difficult or simply unappealing, crowdfunding sites like Kickstarter, Kiva, Indiegogo, Crowdtilt, and Wefunder make wanderers out of all of us. In December 2009, Jesse Genet, a graduate of the Art Center College of Design in Pasadena, asked backers on Kickstarter to finance a project called Lumi, a printing system based on a new technique that allows you to snap a photo and use a smartphone app to turn it into a negative that can be screened onto your T-shirt, jeans, and other objects. The next iteration of the Lumi Process, released in mid-July of 2012, which allows you to print on fabric using sunlight, has also been wildly successful. Genet and her business partner Stéphan Angoulvant needed $50,000 to create kits for anyone who wanted to experience this technology. Their "wanderer" came in the form of 3,525 people who pledged $268,437 to bring the project to life. Welcome to the socialization of wandering.

While so much of Creative Intelligence hinges on our social interactions and relationships, pivoting from creativity to the creation of things takes us out of ourselves more than any other. The competencies of Knowing, Framing, and Playing take us down the path of creativity, and Making involves combining preexisting skills and new tools to create things we might have never imagined possible. But then we often need someone outside our circle, our playground, to provide the resources to complete the journey.

Often wanderers have formal titles, but other times they don't. They can be people with access to capital, networks, and markets that lie outside their usual job categories or careers. Or they can simply be the people who encourage us to keep going. Your pivot circle may start small, with family and friends, and widen to include new partnerships and networks. And it may change dramatically as you scale.

What's important is that you start looking.

BUILDING A PIVOT CIRCLE

Eddie Huang was a twenty-three-year-old law student, living out his immigrant parents' dream of having their son become a successful professional. But his own dream was bigger. During his childhood in Florida, Huang had two big loves: hip-hop music and Taiwanese street food. As he grew up, Eddie became aware of the ways that people connect with their own and other cultures through food. He witnessed people expressing their passions and prejudices in what they listen to, what they wear, what they consume. He wanted to connect the music he grew up with to the food his grandparents had brought over from Taiwan, the small meat-filled buns called *gua bao*, and use this connection as a way to challenge perceptions of immigrant culture in his family's adopted home country. Marry the music of his generation to the traditional Chinese street food of his grandparents, and frame it as critique: Crazy, right?

Huang wanted to call the restaurant Baohaus—a play on the German Bauhaus architecture and aesthetic movement of the early twentieth century. But how to pivot from this idea to launching an actual restaurant—or, more grandly, how to launch a Baohaus brand with a network of restaurants?

Huang did what a huge percentage of start-up founders do; he turned to his family circle. After drawing up a business plan, he went to his father, who'd managed restaurants for years (though never a Chinese one) and he showed it to his brother, Evan, who was majoring in sociology and marketing at the University of Central Florida. Their response? Not wildly supportive. Evan said the idea was original, but further details needed to be worked out for Baohaus to succeed. He said he would help, and jumped on board, thinking they'd get things rolling properly.

Not so for Eddie's parents. Eddie's father may have been in the restaurant business, but he and Eddie's mother wanted their children to do "better" than they did. And so they refused to support his start-up. Luckily, several of Eddie and Evan's aunts and cousins were willing to bet on their idea. But where do you launch such a high-concept restaurant? Not Florida, they knew. On the East Coast, it had to be New York.

THEY ARRIVED IN 2009—A BUST year for the economy, but one that also turned out to be an extraordinary year to launch new companies in the city. As Evan and Eddie checked out neighborhoods, there were FOR RENT signs on storefronts everywhere. They chose the Lower East Side, with its own hip confection of artists, media, and marketing people, plenty of young people—and proximity to Chinatown. In fact, they set up shop in a four-hundred-square-foot space on Rivington Street, not far from the headquarters of another start-up with a musical mission at its core: Kickstarter. Evan's two-week assist stretched on; in fact, he never did go back to school in Florida, enrolling instead in classes on sustainability at The New School.

Baohaus catered the Kickstarter launch party, and the start-up's employees and network were among the first of what Evan calls Baohaus "fans." Not customers, but loyal fans. The circle of people that helped Eddie pivot from creative concept to the actual creation of his brand began with family and friends and expanded to neighborhood people who loved the idea of good street music meeting good street food. Of course, it didn't hurt to have neighbors who were in the process of launching a game-changing start-up.

For about $4, the right price in a downturned economy, people could feast on soft, fluffy buns steamed in lotus leaves, including the signature Chairman Bao, named, of course, after *the* chairman, Mao, who loved the "red-cooked" Mandarin-style meat. The Chairman Bao is made from high-end Niman Ranch pork belly: flash-fried; simmered in rice wine, soy, ginger, rock sugar, star anise (plus Cherry Coca-Cola, which adds a hint of caramel); topped with peanuts; and tossed with red sugar, pickled mustard greens, and cilantro. With offerings like the vegetarian Uncle Jesse Bao, named for a friend, not the character on *Full House*, the menu is constantly evolving.

So is Eddie's pivot circle. Eddie and Evan use social media— Tumblr, Facebook, Twitter, Foursquare—to extend their network of fans and spread their message. Evan even hired one woman who ate at Baohaus so frequently she'd been awarded the badge of "mayor" on Foursquare. The original Rivington Street shop has closed and a foray into a larger Chinese food restaurant did not succeed, but Eddie and Evan are expanding their Fourteenth Street location and planning to open more restaurants. They are also working on scaling the brand by widening their circle of contacts and expanding Baohaus. Eddie has worked with fashion designers and launched his own line of clothing, Hoodman Clothing, that features illustrations criticizing gentrification and other trends that Huang believes are destroying the Lower East Side. On his blog *Fresh Off the Boat*, and on Twitter and Facebook, he is outspoken about "hipster chefs," the role of food in transforming culture, and the Chinese immigrant experience in America. The Huangs have networked with major radio, TV, and online shows to talk about baos, culture, and the politics of food. Eddie has been a guest on *Martha Stewart Living Radio*, offering advice on becoming a chef, and he appeared as the host of a special on the Cooking Channel called *Cheap Bites*. He also has a memoir due out in January from Random House and a new show on Vice.com.

Ask venture capitalists and they will tell you that what Evan and Eddie are doing is very similar to what successful high-tech start-up founders do. They set up shop in a bustling area congested with people launching their own businesses. They cultivated a network of people who can help them scale their creativity. They didn't limit themselves to one platform, looking instead for natural partners in other industries who would "get" them.

No matter your field, your pivot network is key to getting your idea into the world. One incubator that nurtures young entrepreneurs and helps them develop their ideas, Y Combinator, was founded on the premise that an entrepreneur's original idea may not even matter when it comes to building a new company. The important thing is the circle of people who surround the founders, providing the scaling skills and capital necessary for pivoting.

So how do you build your own pivot network?

Pivoting Organically

Starting with those close to you, whether people inside your large organization or your family and friends, is perhaps the most common first step in moving from creativity to creation. The first customers of most architects, artists, and start-up founders, for example, are often their parents or friends. Architect Charles Gwathmey was known for his contributions to the style of High Modernism, designing a number of high-profile commercial buildings as well as residential buildings for clients including Steven Spielberg and Jerry Seinfeld. But his first building, an unusual round summer cottage and studio in Amagansett, New York, was for his parents.

In cases like Eddie Huang's, starting with family members might be enough. His dad's restaurant experience, his brother's business management training, and investment from other family members were crucial in the launch of Baohaus. But at a certain point, Huang needed to move beyond his small circle, and he did so with the help of people in fashion, the media, and the wider world of food culture.

Once you've looked to your circle of friends and family for advice, support, and investment, with certain companies it can make sense to "rent" space on a sales platform such as Etsy and Amazon and eBay. There are also manufacturing platforms around the world, OEMs (original equipment manufacturers), that make it easy to scale. Pivoting by renting space on a platform is allowing thousands of people to keep control of their creations while scaling them nationally and globally.

PIVOTING WITH THE HELP OF LOCAL NETWORKS

Investors and other wanderers tend to group around cities and universities. They can be found right in your own company, of course, or among your circle of friends; many young entrepreneurs find their partners in their dorm rooms or right down the hall. Universities are excellent places to begin building a pivot network—a growing number are even providing capital and marketing advice to their students through venture funds consisting of their own successful alumni.

Going to networking or start-up events in your city—here's where being in the "right" city can really pay off—is a great way to meet people who can help you make your idea a reality. NY Creative Interns, started by Emily Miethner, puts on events for young graduates all the time. Past events have included speakers from NBCUniversal, Bravo TV, Mediabistro, Etsy, Google, Condé Nast, and Christie's who've talked and mingled with the audience.

Pivoting is a way to leverage your creativity through the kindness, expertise, and capital of family, friends, and strangers. Your goal is to take your original creation and find others who see value in it and are willing to supply resources to bring it to life. In the past, this meant networking among a relatively small number of gatekeepers to these resources. Thanks to social media, that number is vastly larger—but much easier to access. You don't need to be "connected" anymore, although it certainly helps. You can crowdfund your ideas and get them in front of millions who might be willing participants in your dream.

Pivoting by Linking Up with Another Company or Platform

Pivoting is a Rorschach test of your entrepreneurial self. Moving from creativity to capitalism will always require some level of scaling up, but there is more than one way to increase the size of your start-up. There are, in effect, two kinds of pivots and two types of pivoters. Knowing which kind of pivoter you are is key to your creative success. Ask yourself: Do you love the early stages of creation—the "whiteboard" moment, the prototyping process— more than the idea itself? Or do you want to go further with a particular idea, nurturing and growing it into a big, dynamic business or nonprofit?

Many creators are "serial entrepreneurs" who go from one creation to another, leaving it to others to develop and grow each idea. Other creators are "entrepreneur builders," who transform themselves into leaders and managers of big organizations as they build out their singular creative ideas. Understanding yourself— your true desires and your real capabilities—is an important part of pivoting. Deciding how far you want to go with each idea is a key personal and business strategy.

YouTube cofounder Chad Hurley found himself in this position not long after launching his start-up. He needed to grow, but couldn't do it alone. In a sense, his wanderer came in the form of a multibillion company, itself a start-up less than a decade old.

Hurley was born into the world of social media. He went to Indiana University of Pennsylvania in the nineties as a computer science major but switched to graphic design with a minor in print-making. He figured that computers and code were a means to an end, not an end in themselves. What he really wanted to do was design great interfaces online that allowed people to engage and interact in an easy, pleasurable way. "I wanted to use computers as a tool, not make them," he said, in an interview with Bill Moggridge.

Hurley's first big job as a designer out of school was to come up with an easy way for people to pay online. We take for granted how easy it is today, but back in the nineties, it was a nightmare. While working at the PayPal division of eBay, Hurley worked closely with another young designer, Steve Chen, and they came up with a simple solution—a big button. There are now buttons on nearly every online site that allow you to pay now, donate money, move to a shopping cart, and more.

Hurley got his start designing at the dawn of the era of social media; Friendster and MySpace, the forerunners of Facebook, were already popular. Caterina Fake had already launched Flickr, allowing millions of people to post their pictures and images online and share them with their friends. But no one had yet designed a way for people to upload their videos and share them. There were video sites in existence, "but they didn't let the community post their videos," says Hurley. "We wanted to give people the power to deal with their own video."

Hurley left PayPal and took his buddies Steve Chen and Jawed Karim with him to start a company that would do just that. They successfully built out the company to sixty-seven people, but there were big issues with advertising, maintaining servers, and dealing with big media companies. Hurley was looking to expand the community of video makers and sharers that he was part of, but there were constraints on his ability to do it.

And so, Hurley and his cofounders turned to people who were members of his generation but not necessarily the culture of social media to which they belonged—Google. Larry Page and Sergey Brin, founders of Google, were all about the mathematical algorithms of search, not the emotional connections of sharing and community. They were wildly successful at enabling individuals to search for information and had built a multibillion-dollar corporation based on it. But whenever Google tried to do social, it had trouble. The company had tried to build Google Video, but it was too complex. And Google had asked people to pay for the service—free on YouTube.

So Hurley sold YouTube to Google for $1.65 billion in 2006, only eighteen months after he launched, thereby freeing himself up to pursue other projects. Pivoting by linking to a bigger platform—YouTube's sale to Google, Kevin Systrom and Mike Krieger's sale of Instagram to Facebook—is perhaps the strategy we've become most familiar with in the era of start-ups.

Of course, the strategy works the other way too—rather than selling to a larger platform, some entrepreneurs scale by expanding their original platform. Brian Chesky, cofounder of Airbnb, bought smaller platforms similar to his company's model—Crashpadder and Accoleo, both in Europe. This has been Larry Ellison's strategy at Oracle as well. And eBay has grown by buying PayPal and other smaller companies. Even Apple has begun to do this by buying the company that made Siri.

Of course, you can still pivot and decide *not* to scale, at least not much. The growing local movement is putting a higher value on staying in the neighborhood than it had in the era when globalization was uncritically celebrated. Besides, there are many platforms today such as eBay or Etsy that allow you to market globally while staying put.

BUILDING THE RIGHT KIND OF pivot circle is essential when you're still in the "idea" stage, when you've begun prototyping or testing your idea but need feedback, or when you've got the business in motion but need help growing. Not only can this trusted circle of advisors and partners help you build upon your creativity, inspiring you to explore areas or think in ways you might not have had before, it is indispensable when it comes to transforming your idea into something tangible and, in the case of business, marketable.

But what if you're a "newbie" and haven't ever created anything, let alone a business? The idea of reaching out to a network of strangers, or even tapping your family and friends, can seem daunting. Chasing after contacts requires a level of confidence you may not yet have.

Yet you probably have a passion about something that so animates you it gets other people excited too. You know something so well or are interested in something so new or simply see something old in such a totally different way that people are drawn to you. You may not consider your enthusiasm a sign of a "calling" but that's what it is. You may not see your friends and colleagues who find meaning in what you're interested in as a "following," but they are. And you may not think you can harness that enthusiasm to become more charismatic, but you can.

THE POWER OF CHARISMA

On January 28, 2010, the *Economist* featured a religious image on its cover. A man in robes stood smiling, beams of light emanating from a sun behind his head, a tablet in his hands. The man was Steve Jobs, the "tablet" he was offering humanity was the iPad, and the message was clear: On the cover of one of the world's most important business magazines was the Prophet of Profits. The text above the picture read "The Book of Jobs," clearly framing the Apple cofounder and CEO as a religious figure engaged in a powerful relationship with his acolytes and believers.

I imagine the *Economist* chose the religious metaphor, in part, because it's one that's familiar. But Walter Isaacson, too, described Jobs using language typically reserved for religious leaders, arguing that his "absolutism, the ecclesiastical bearing, the sense of his relationship with the sacred, really works."

Look around a bit and you'll notice that many of the new ideas and products that are transforming our lives were brought to us by entrepreneurs who are, in their unique ways, charismatic. Charisma, secularized from its religious roots—it comes from the Greek χρισμα, meaning "gift of grace," or "divine favor"—is a powerful force responsible for the creation of some of the most important changes in our economic lives. Yet in the rush to build processes of innovation in our organizations, individual charisma has been largely ignored as a driving force for creativity, the creation of new products, competition, and capitalism itself. Charisma is not central to any economic model, and yet it is often what helps innovators bring their idea beyond the moment of conception and into the lives of hundreds, thousands, millions of people. And it's something we might overlook if we're content with the standard definition of charisma.

Many people assume charisma is a state of being. In this way it's a lot like our views of creative genius—either you're charismatic or you're not. We see the charismatic person as imbued with an internal light (or in some cases darkness), that rare person who "has" it because he or she was born with it. The rest of us can do little but follow.

But that view of charisma—one that focuses solely on the individual and not the community that gets behind him or her—doesn't hold much water at a time when even the shiest or geekiest among us can make YouTube video that attracts a massive audience of followers, when leaders of even the largest organizations are held up to constant scrutiny by anyone with a Twitter account. And perhaps that definition never really did hold up; where, after all, would the charismatic leader be without loyal followers?

IN *The Protestant Ethic and the Spirit of Capitalism*, Max Weber argues that there are three forms of authority. There is power that comes from tradition, such as the power of parents over children. There is power that stems from the legal and rational authority that flows from organizational structures, such as the role of CEO or the President of the United States or the pope. And then there is the power that comes from the "quality of an individual personality by virtue of which he is set apart from ordinary men," according to Weber. "The holder of charisma seizes the task that is adequate for him and demands obedience and a following by virtue of his mission."

Weber argues that charismatic leaders tend to arise in times of social stress. They challenge established bureaucracies and the normal routines of everyday life. Perhaps most important, charismatics challenge the established economic order and the orthodoxies of conventional business. "In contrast to any kind of bureaucratic organization of offices," wrote Weber, "the charismatic structure knows nothing of a . . . regulated 'career,' 'advancement,' 'salary,' or regulated and expert training of the holder of charisma or of his aids."

In other words, the charismatic sounds a lot like an entrepreneur.

Charisma Is a Calling

When Facebook "went public" in May 2012, eight years after it started up, it was valued at over $100 billion and Mark Zuckerberg's personal fortune (on paper) soared to about $19 billion. But in all the conversation around this enormous financial transaction, there was nary a word from Zuckerberg about shareholders or profits. In his IPO letter to potential shareholders, Zuckerberg said, "We believe that a more open world is a better world because people with more information can make better decisions and have a greater impact. That goes for running our company as well." To ensure that sentiment, Zuckerberg arranged for a special stock arrangement enabling him to continue controlling Facebook no matter what the public owned.

We're inclined to dismiss vision statements of corporations as nothing more than public relations, but the statements of entrepreneurial founders should be taken much more seriously. They are often more than PR pablum and represent strong beliefs that the founders actually act on. Most important, they often reveal a great deal about a founder's motivation for creating.

To my mind, Zuckerberg was revealing his belief in his "calling" to make society more open. This drive to do something more than simply generate profits is common to most charismatic entrepreneurs. Money, certainly, is a factor, but either as a means to continue implementing the dream of the calling or as a marker, to show other charismatics how well you are doing. Of course, the subsequent IPO disaster, with the initial offering price of $38 plummeting to half, left many individual Facebook investors angry and feeling betrayed by Zuckerberg. The closed nature of the IPO, with insiders selling out at high prices to small investors who then lost money, eroded Zuckerberg's message. If he is to retain the loyalty of his following and keep them engaged with Facebook, Zuckerberg has to prove his calling is not primarily about money but about meaning. He took one step in that direction in September 2012 by announcing he would not sell his stock and cash in for at least another year.

A secularized version of a calling has long been associated with capitalism. From the 1920s through much of the 80s, when professional managers ran most of America's large corporations, business leaders saw themselves as professionals serving a broad range of interests, many of them social. They felt a collective responsibility to stakeholders—employees, local communities, the national government, customers, suppliers—as well as shareholders. In the 1990s, the CEO's role was recast as a maximizer of shareholder values, but before that "a higher interest was the sin qua non of business professionalism," says Harvard Business School professor Rakesh Khurana. The heads of big corporations felt they had a "calling" to do good for the nation.

This sense of calling is now rare among CEOs of global corporations, who focus on shareholders and see themselves as global citizens, not leaders of local communities. CEOs often have international responsibilities that surpass local and even national obligations. But the idea of a calling is alive and well among the founders of start-ups like Zipcar or Method who often embrace a social challenge with an entrepreneurial solution. The notion of creating neighborhood jobs and sourcing materials locally is strongest today among new companies growing up, rather than big companies going global.

Back in 2004, when Sergey Brin and Larry Page were taking the company they founded public, they, too, expressed a "calling" for their effort. "Sergey and I founded Google because we believed we could provide an important service to the world—instantly delivering relevant information on virtually any topic," wrote Page. "In pursuing this goal, we may do things that we believe have a positive impact on the world, even if the near-term financial returns are not obvious."

Charismatic leaders understand that the relationship between a leader and a community involves an exchange. What binds charismatic leaders to their followers is the promise of a gift, an unexpected surprise or powerful tool, that gives meaning to their lives. Charisma is a quid pro quo. If the gifts stop coming, if they lose their meaning, the allegiance of the following dies and the relationship ends. Think of all the great writers and artists who lost their "glow" when their work faded and their audiences disappeared.

Charisma starts with your calling, but how you present that calling is a different story. It often takes months, years to develop the confidence to present your ideas and yourself in a way that truly grabs people and makes them want to get involved. Once you clarify your passion, your next step is to communicate that passion effectively.

Your favorite actor, your best teacher, your impressive friend who just raised thousands of dollars on Kickstarter or persuaded a VC to invest in her company, all have some amount of charisma. But most of them probably weren't born that way. They learned how to be charismatic, and so can you. The challenge is to find your own personal kind of charisma.

Charisma Can Be Learned

Though Steve Jobs is now considered one of the most charismatic CEOs of the decade, he did not start out that way. Jobs progressively increased his level of charisma over the years, choosing a signature style of personal dress, the simple black turtleneck; learning to introduce products with Broadway-level drama, whipping off a cloth to reveal a new offering and promoting what Apple employees called his "reality distortion field," demanding even that which appears to be impossible.

I saw Mark Zuckerberg for the first time at the World Economic Forum in Davos in 2009. He was sitting in front of several hundred people in a large conference room on a panel about the future of mobile along with Chad Hurley and a number of other high-tech luminaries. Panel moderator Mike Arrington, the founder of TechCrunch, said it was the first time he ever saw Zuckerberg with a tie on. Zuckerberg responded, jokingly, "No, I wore one all through boarding school." Over the next forty minutes or so, he smiled, laughed, and talked easily about Facebook, privacy, and his goal of having a more open society. He was, in a word, charismatic.

But he didn't start out that way.

Mark Zuckerberg was once seen as a strange young man who took an unusually long time responding to people when asked a question. But charisma requires delivering "secular epiphanies" again and again to your followers—and that means learning how to manage a complex series of relationships and information. You need to find people who can teach you. Successful entrepreneurs are adept at using their pivot networks as mentoring schools.

Zuckerberg is a master at finding teachers, mentors, and partners—and also at knowing when to part ways with them, as Henry Blodget detailed in a 2012 article in *New York* magazine. Early on, he enlisted the help of Sean Parker, cofounder of Napster, who connected him to Silicon Valley VCs like Andreessen Horowitz, but when Parker's "party boy" persona was deemed a "liability," Zuckerberg dropped him. With enough money to do a start-up, Zuckerberg hired Owen Van Natta from Amazon, and

Van Natta was instrumental in increasing revenue and building out the company from twenty-six employees to hundreds. Then he fired Van Natta and hired Sheryl Sandberg, a veteran manager from Google, to help him scale Facebook to a global corporation. Along this journey, according to Blodget, Zuckerberg sought counsel from the likes of Peter Thiel, who was an early investor; Marc Andreessen, now a board member; and LinkedIn's Reid Hoffman.

None of this was easy for Zuckerberg, who was far more comfortable as a software programmer and product designer. But he pushed himself to find people who could teach him a wide variety of communication and business skills, which he used to further his vision. When I saw him in Davos, he was already on his way to shape shifting from computer nerd to high-tech entre- preneur, exuding a confidence in his calling that didn't exist when he was in school. Today Zuckerberg, a multibillionaire, is one of the most charismatic people in business. His learned charisma was an essential tool in pivoting his tiny social media company— started years after MySpace and Friendster launched the concept and each built up an impressive number of users—into one of the largest companies in the world.

PIVOTING: TURNING YOUR IDEA INTO A GAME-CHANGER

No two people will have the same pivoting strategy. For some, pivoting might begin the moment you decide to go from knit- ting scarves for your family to setting up an online storefront. For others, it can happen when you get enough funding to hire thirty more people to join your start-up. For still others, it may mean finding the person at your company with enough clout to make your dream a reality. But while the precise pivoting strategy will differ from person to person, the essential ingredients remain the same: You need a product with aura, you need a wanderer to help you bring your idea to the wider world, you need to build a pivot network and decide over time how to nurture and grow that network, and you need to cultivate charisma by learning how to clearly articulate your calling.

And so your first challenge is to figure out where in the pivoting process you are. Are you still in the idea phase? If so, what kind of feeling do you want your creation to inspire in people? Where might you go for inspiration? The story of the creation of the iMac reveals how something as simple as a jelly bean can inspire in surprising ways. Your own journey toward imbuing the things you create with a deeper feeling or meaning can involve taking a trip to the country as the leaves are changing, remembering a toy you loved as a kid, or listening to a forgotten style of music. But that initial inspiration is only part of the magic of aura. Ives and his team didn't just think of the jelly bean as visual inspiration and leave it at that; their investigation into the way the candy was made became a road map for the entire manufacturing process.

Throughout the pivoting process—long before you think about scaling and long after you've bought your product to market—the importance of cultivating charisma cannot be underestimated. There are few classes at business, art, or design schools where students can go to learn how to be charismatic. It's a quality that you're supposed to be born with, something you inherit, a quality you either have or don't. But there are dozens and dozens of examples of people who over the course of time became charismatic.

So how did they do it? More important, how do you?

You can begin by asking yourself what you're passionate

about. We all have a passion, something that we love. It's often off to the periphery of our jobs or our lives. It's that secret hobby or that thing you used to be really good at, or that new approach at work that you feel would be so much better than what everyone is doing now. It's that idea that you have about a product that your friends would really love to try if they could.

Much has been written about the importance of following your passion, but anyone who's juggling life's many demands—moving up the career ladder, raising children, making ends meet in a tough economy—might wonder how to find the time. But a quick glance on Kickstarter will tell you that following your passion to create something can be a very smart investment.

Moreover, that thing that electrifies you, that you can't stop talking or thinking about, is what can draw an audience to you. When we talk about what we love, our faces light up, the words flow, and the excitement can be contagious. What animates you can animate others and make them want to get involved. Once you identify what excites you, start talking about it. If the first person you talk to isn't interested, keep talking until you find someone who is. Start in your local community, find people interested in similar ideas online. Seek out mentors and "pivot network" members with different areas of expertise. Who do you know who's great at logistics? Who seems to know *everyone* and might put you in touch with people who can help make your idea a reality? Most important, who gets you excited? Who's supportive and encouraging? We all have that friend (or hopefully more than just one) who makes us feel energized and invigorated; even if he or she doesn't have "traditional" business expertise, this kind of support cannot be underestimated, and this kind of person may be an invaluable member of your pivot network.

Your pivot network will vary depending on your current position and goals. Are you an artist or designer looking to take over more control of the creation and distribution of your product? Are you a manager charged with finding the next big thing for your company? Where do you need help right now—can you bring in others who excel in areas where you lack experience?

It's also helpful to recognize which side of the pivot you lean toward. Do you love coming up with ideas and bringing people together, do you thrive on the enthusiasm and excitement of the whiteboard stage? Or do you like building things, linking people from different networks, managing day-to-day operations?

Challenge yourself here when answering these questions. Like many people, you may see yourself as an "idea person" who either can't or doesn't want to focus on implementation. But what happens when you can't get enough people behind an idea you really believe in? What if you can't get funding through traditional channels? Don't let the "idea person" label stop you from following through. Having a lot of ideas and enthusiasm at the early stage of a project is incredibly important, but you may not yet see how your experience organizing other areas of your life can be applied to this work. In short, there's nothing wrong with embracing the idea stage of the pivot. But don't write off the other side if you've never given it a try.

On the other side of the spectrum, you may feel more comfortable throwing your support and implementation skills behind other people—perhaps those you consider "more creative." In her classic book on creativity *The Artist's Way*, Julia Cameron wrote about "shadow artists," people who defer their own creative dreams for any number of reasons, working in support roles or even dating artist types as a kind of consolation prize. Of course, support roles can involve work that's every bit as creative, challenging, and rewarding as those traditionally viewed as "artistic" or "creative." Indeed, wanderers are just as essential to the creation process as the person who had the first spark of an idea—and we should be celebrating these people every bit as much as we do those who prefer the whiteboard stage. But it's important to take stock of whether you're putting off personal projects to focus solely on furthering someone else's creative dreams.

Once you understand that charisma is a social relationship, the steps you need to take to make yourself more charismatic become clear. Finding your passion gives you the emotional capital that you need to begin connecting to a larger audience. Learning how to communicate gives you the tools to do that. And reaching out to find mentors to teach you how to scale your message and your audience takes your charisma to an even higher level.

Pivoting should play an important role in any organization's strategy. That means not being afraid to step away from hierarchical models in order to facilitate innovation and giving people more control about how they team up and partner. It means hiring or training in-house wandering managers who can help employees bring their ideas into being. These wanderers should have experience in the design or creation of products, not simply in managing. They need the right blend of vision, discernment, and domain knowledge—an understanding of the company or industry's past and a willingness to drive it into the future.

It's important for individuals and organizations alike to understand that the work of pivoting is never really done. Starting with your circle of friends and reaching out to local networks is a key starting point, but once your business is up and running you'll need to enlist the help of new wanderers with new skills and areas of expertise. You'll need to continually deliver on the promise you've made to the community from your first days of existence. When your product loses its aura—when quality goes down or the brand grows too large and diluted—people notice. When charismatic founders leave, they often take the aura of the product with them, and so it's up to the people who remain to keep the original dream and spirit of creation alive.

Pivoting is the stuff of legend. It is the narrative of your origin myth. How you went from having a little idea to launching a game-changing business or nonprofit is a story we all love to hear. And why is that? Perhaps it is because for all the emotion we feel for creators, it is the builders whom we hold in highest esteem. Those who take our creativity and make it manifest in living, breathing, useful ways are at the apex of our admiration. And rightly so.

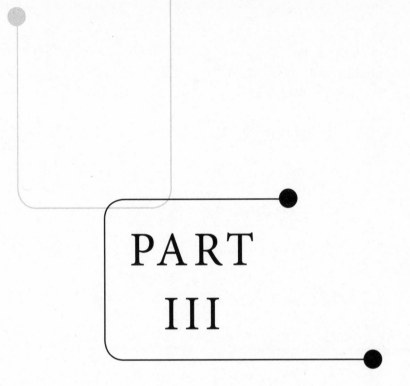

PART
III

*The Economic
Value of
Creativity*

Indie Capitalism

In May and September 2012, Hewlett-Packard announced layoffs totaling 29,000, nearly 8 percent of its entire workforce. The decisions were made under CEO Meg Whitman, the fourth CEO in six years. The innovation giant was floundering—dying, really—sideswiped by changes its management didn't see and couldn't cope with. Where once HP led the world in scientific instruments, then printers, then PCs, it botched the shift to mobile devices—smartphones and tablets—and wound up buying into consulting and more PCs. Where rivals integrated software and hardware, HP relied on Microsoft and Intel for the guts of its machines. Even with a cutting-edge new technology—Halo—that allowed people in San Francisco to conference with others in Beijing almost as if they were sitting in the same room, HP couldn't execute and rival Cisco swept in to capture the market.

How could such a pioneering company fail to innovate? Too easily, as it turns out.

In the spring of 2012, I gathered together six engineers who worked at HP during the height of its success to discuss the culture of the company, and how it had changed over time. They were all men on the cusp of sixty, if not slightly over. After I talked with them for two days, it was clear that their work at HP, their commitment, was not about money; they did it for the fun, the excitement of creating something entirely new. The ThinkJet printer was just one of the many innovations to come out of the autonomous, supportive environment where wandering general managers helped fellow engineers take new products to market. "As long as we could create something that had some sort of margin that allowed it to potentially become a product, we could do it," said one engineer who worked in the HP labs during his thirty-plus years at HP. It was a culture nurtured by the founders long after they began the company out of their garage. "Bill and Dave cared only about your making new things," he said. "That was a 'We can do anything' time," another recalled. "We thought about the future, five to ten years out, and beating the competition."

But things changed. China emerged as a source for cheap manufacturing. A younger generation wanted technology that enabled a more mobile digital life, available whenever, wherever. Wall Street began demanding ever-rising profits every three months. A financial culture that reduced everything to metrics, markets, and monetary transactions spread from Wall Street to Washington, DC, and then to Main Street, from the corner offices to the scientific laboratories of America.

HP, too, embraced a financial culture when Carly Fiorina took over in 1999. Fiorina was brought in by the HP board to change strategy, and that's what she did—by buying Compaq, a PC maker, and merging it with HP's PC business. Her strategy was to shift HP's culture away from organic growth through internal innovation and toward expansion through mergers with outside companies. Problems with the Compaq merger meant HP's managers were preoccupied with integration, not innovation, for years, even after Fiorina left. Fiorina's timing wasn't all that great either. While she was boosting HP's personal computer sales, Apple was getting ready to disrupt the entire PC market with another one of its innovations, the iPad. HP's PC sales slowed once the tablet was introduced in 2010.

The greater damage came from Fiorina's dismantling of the culture of confidence Packard and Hewlett had instilled in the company. She was, as one former HP engineer puts it, a "total control freak"—her style a stark contrast to the empowering environment fostered by the company's founders. Decisions once made by general managers with engineering expertise were kicked up the corporate bureaucratic line to vice presidents with MBAs. She brought in consultants from firms like Bain who advised her on how to make HP more efficient.

The trust between lab engineers, scientists, and their GMs had been broken and the line that connected creativity to new products severed. New ideas were thereafter judged through the lens of data-driven efficiency analytics, not the "we can do anything" culture of Hewlett and Packard. Probability replaced possibility as a criterion of acceptance. The HP Labs began to dry up.

Fiorina was pushed out after five years, replaced by a cost-cutting operations expert, Mark Hurd. He outsourced much of HP's manufacturing, according to one of the former HP engineers I talked to, weakening the ability of engineers to do what they did best: innovate. He cut spending for R&D to pump up short-term profits. And, in what was perhaps the most destructive practice he instituted, the former HPer said that Hurd required all managers, including lab chiefs, to automatically fire the "bottom" 5 percent of their staff every year. Straight out of a management consulting handbook, this practice was supposed to improve the quality of human capital year after year. What it did was destroy the trust people had in one another. Engineers now had to compete against one another; their projects became bricks thrown at one another, not gifts created and offered up in the hope of making some new and awesome product. "Once you started worrying about your job, you started getting parochial, keeping the best things to yourself," said one of the engineers. "You're not focusing on what's the best thing to do."

Over the past decade or so, HP has struggled to cope with a sea of forces smashing up against its business operations. But instead of looking to the creative abilities of its labs and engineers to meet the challenges, the board of directors chose mostly outside CEOs. That's a strategy that can work when a leader understands the aspirations of the company's customers and guides employees toward innovating new ways to meet them. But time after time, the CEO brought into HP to deal with the vast changes sweeping the economy could offer only what he or she already knew— centralize, cut, merge, outsource.

Outsourcing was particularly destructive. After management consultants advised HP to save money by outsourcing engineering jobs to India and Eastern Europe, people in the HP labs began to lose connection with members of their team overseas. "We would see them only once a quarter and do global conferences on the phone half asleep," said an ex-HPer. "There was no contact and you would miss that interaction. Innovation just explodes when you talk face-to-face," he said. And fizzles out when you don't.

HP fights to remain number one in PC sales worldwide, but Beijing-based Lenovo is fast closing in. In the third quarter of 2012, Lenovo even surpassed HP's global market share, shipping 15.7 percent of PCs against HP's 15.5 percent in that period, according to the research group Gartner. The company brought out a tablet to compete with the iPad, only to withdraw it in months. A huge surge in 3-D printing is sweeping the United States and Europe, yet HP is scarcely to be seen in this market. Even when its labs produced a truly disruptive innovation, the Halo conferencing technology, HP didn't manage to spin it into a successful product, ultimately selling it to Polycom.

Many companies have fallen into HP's trap of managing for efficiency, not creativity. Whole domestic industries, from steel to consumer electronics, have been gutted as a result of focusing on the short-term financial benefits resulting from efficiencies of scale at the expense of long-term investment in internal innovation. CEOs and top managers trained in metrics and analytics understand the process of squeezing more and more profits out of existing products. What they don't get is that the profits from innovative new products can have greater value, support higher prices, and generate even greater profits.

THE RISE OF FINANCIAL CAPITALISM

I attended the World Economic Forum in 2008, just months before the collapse of Lehman Brothers. Hedge fund managers had taken over the tiny village of Davos, Switzerland. They were booking the best hotel rooms, throwing the biggest parties, running the most important panels, and squiring the most beautiful models in the fanciest cars. That year, the influence of hedge fund managers eclipsed that of the old global elite of high-tech hot-shots, corporate CEOs, and presidents and premiers from around the world. Even though Facebook's Mark Zuckerberg was there, along with U2's Bono and British Prime Minister Tony Blair, it was the hedge fund crowd that caused the buzz.

I was witnessing evidence of a shift in power so dramatic that it could be characterized by what sociologist Erving Goffman called a "status bloodbath." It had taken decades, but the financial types had triumphed over the builders and creators of real things, as well as the politicians who ostensibly ran the world. And even though the Great Recession would soon raise doubts about financial capitalism and its corporate correlate, shareholder capitalism, they have remained the dominant economic models in America.

I may have witnessed the apogee of financial capitalism in Davos, but it got its start some four decades earlier in Chicago.

In May of 1970, Eugene Fama, a professor at the Booth School of Business, published an article in the *Journal of Finance* called "Efficient Capital Markets: A Review of Theory and Empirical Work." In it, Fama would take Adam Smith's theory of the "invisible hand" to new levels.

Along with that of Milton Friedman, Frank Knight, George Stigler, and other economists at Chicago, Fama's work led to the economic model we now call the efficient market hypothesis or efficient market theory. In its purely financial form, EMT attempted to describe how stocks and markets functioned. Like all complex economic models, it relied on a handful of basic assumptions:

1. Rationality: people act rationally when making economic decisions;
2. Efficiency: markets are always efficient because they accurately reflect the information available at the time;
3. Equilibrium: markets tend toward balance and equilibrium;
4. Risk: managing risk efficiently is the key to maximizing profits;
5. Measurement: only what can be measured should be included in the model.

Of course, what was missing from the efficient market theory was as important as what was included: uncertainty. Because the theory constituted extreme and unpredictable occurrences of the kind economist Nassim Nicholas Taleb called "black swans" in his 2007 book of that name, the Chicago economists viewed uncertainty as an "exogenous" variable. By excluding uncertainty and focusing on measurable risk (it was Chicago economist Frank

Knight who introduced the distinction between risk and uncertainty), the efficient market theory model of economics assumed and reinforced a culture of control.

This would have important consequences for creativity, innovation, and economic growth over the next decades. It is, after all, uncertainty that forces us to look for opportunities to create new things that have new economic value. Entrepreneurs find new opportunities within the uncertainties of new technologies and changing social relationships and values. Like artists, they thrive within a culture of chance, not a culture of control.

It's not as if economists had never seen proof of the effects of uncertainty: There is a significant literature on market panics going back over a century. In the 1960s and 1970s, as EMT was being developed, Hyman Minsky was writing about booms leading to euphoria, panic, and busts. Charles Kindleberger's *Manias, Panics and Crashes: A History of Financial Crises*, was hugely popular when it came out in 1978. British journalist and essayist Walter Bagehot had written about financial crises in Lombard Street, London's financial district, back in the latter half of the nineteenth century. And who hasn't heard of the extraordinary tulip mania in the Netherlands of the 1630s, popularized in Charles Mackay's 1841 book *Extraordinary Popular Delusions and the Madness of Crowds?*

Of course, we don't need to look that far back, considering that we've all lived through two major manias of our own, the dot-com bubble and the housing boom and bust. It would appear that uncertainty with all its promise and perils is very much a part of our economic system. Which makes its absence in our economic models all the more troubling.

An economic model based on rationality, predictability, and measureable risk promised much more to bankers, corporate managers, and policy makers than one based on uncertainty. If risk could be measured and managed, then it could be controlled. If risk could be controlled, then it could be increased and leveraged. For bankers, an efficient market theory was a persuasive argument in favor of wildly leveraging their capital to generate profits. Risk, after all, was under control. For corporate managers, EMT provided a rationale for controlling complex organizations at a time when "just-in-time" supply chain management was making it possible to expand globally. And so business schools began to focus on developing mathematical tools that aided in efficient management.

The belief in the efficient market was so powerful that in 1976, economists Michael Jensen and William Meckling connected it directly to CEO compensation. In their work, including a paper that is often cited as ushering in the era of financial capitalism called "Theory of the Firm: Managerial Behavior, Agency Costs and Ownership Structure," published in the *Journal of Financial Economics*, they argued that if the interests of the manager and shareholder were identical, they should be compensated by the same thing, the company's stock price. CEOs should be paid mostly with stock and stock options and run the company for one purpose, to maximize that share price. In other words, CEOs should focus less on stakeholders such as employees, suppliers, local communities, and national governments because doing so compromised their ability to extract the maximum amount of profit. It wasn't efficient.

By the mid-eighties, boards of directors tied the compensation of CEOs directly to the share price of the companies they managed. Executives were offered millions of stock options to hit stock market targets. Roberto Goizueta, the CEO of Coca-Cola from 1981 to 1997, became the first manager in America to become a billionaire from a company he had neither founded nor taken public.

Financial capitalism ushered in a new business model akin

to a global game of efficiently squeezing costs and maximizing short-term financial profits. And despite the worst recession since the depression, we remain frozen in that model. Top managers, locked into the stock price of their companies, are expected to meet or exceed the quarterly estimates of Wall Street analysts. "Shareholder value" is the paramount, often the only, guiding principle to corporate behavior, with stock prices on financial markets the one signal of success or failure. After talking to Wall Street recruits while doing fieldwork for her book *Liquidated: An Ethnography of Wall Street*, Karen Ho, a professor of anthropology at the University of Minnesota, wrote that "shareholder value was the most important concept with which my informants made sense of the world and their place in it: it shaped how they used their 'smartness' and explained the purpose of their hard work. . . . Creating shareholder value was morally and economically the right thing to do."

THE FINANCIAL CRASH THAT BROUGHT on the worst recession since the Depression has, of course, tarnished the efficient market theory. Throughout the economics establishment, there is recognition that the model didn't work as promised. First, the markets didn't act rationally. Losses in one small sector of the huge, multitrillion dollar financial market—subprime mortgages—were enough to set off a worldwide panic, leading to the collapse or near-collapse of some of the largest financial institutions. Second, markets didn't behave efficiently in 2008 and have yet to return to the balance and equilibrium EMT promised. As I write in 2012, there are still huge imbalances in the capital, labor, and trade markets. The value of millions of houses remains less than the mortgages people took out to buy them. Interest rates, zero for many financial instruments, are below inflation. Finally, and perhaps most important, the volatility of the markets during the crash did not reflect the underlying economic fundamentals but rather reflected the heightened emotions of political and social actors.

It took years for economists, policy makers, and bankers to admit that the fundamental assumptions underlying the efficient market theory were wrong. In a 2010 interview with Martin Wolf, the economics correspondent for the *Financial Times*, Lawrence Summers, former Treasury Secretary under President Clinton and ex-assistant for economic policy for Barack Obama, grudgingly admitted he was surprised at the failure of efficient markets in the crisis. Summers was, after all, a chief architect of the deregulation that helped provoke the financial crash in 2007. In 1999, Summers, along with Treasury Secretary and ex-Goldman Sachs co-CEO Robert Rubin, did away with Glass-Steagall, the Depression-era regulation of the financial markets, calling the repeal "historic legislation" that will "better enable American companies to compete in the new economy."

Perhaps no one believed more in the rationality and efficiency of markets than the former chairman of the Federal Reserve. But in October 2008, Greenspan told a congressional committee that he had put too much faith in the self-correcting powers of the markets. "Those of us who have looked to the self-interest of lending institutions to protect shareholders' equity, myself included, are in a state of shocked disbelief," he told the House Committee on Oversight and Government Reform, as reported in the *New York Times*. Greenspan was criticized for opposing greater oversight of the trillion-dollar subprime mortgage market in credit default swaps, which were supposed to reduce risk but were misused to leverage risk to historic levels. In response, he said that "this modern risk-management paradigm held sway for decades." But "the whole intellectual edifice, however, collapsed in the summer of last year."

By the middle of 2012, even the bankers themselves were

admitting the model was a failure. Back in the mid-nineties when Sandy Weill became CEO and Chairman of Citigroup, he lobbied to end Glass-Steagall because he believed it would lead to bigger, more efficient, more competitive banks. He then combined insurance companies, commercial banks, and investment banks into one giant bank, Citibank. During the financial crash, it took billions of dollars from the federal government to save it. When efficiency, rationality, and equilibrium all failed, taxpayers had to come to Citibank's rescue. On July 25, 2012, Weill said on CNBC that regulation was needed again for banking and that Glass-Steagall should be put back into place. "What we should probably do is go and split up investment banking from banking," he said. "Have banks do something that's not going to risk the taxpayer dollars, that's not going to be too big to fail." Not only was the outcome bad for Citi's shareholders, it was bad for the entire nation.

While economists and policy makers engage in discussions about the failure of the efficient market theory in finance, little is being said about the impact the model continues to have on the business world. Here, the damage may have been even deeper and longer-lasting. At its most obvious, the theory has not delivered the corporate profits it promised. According to Rotman School of Management Dean Roger Martin, writing in the *Harvard Business Review*, from 1977 to 2008, the years when financial capitalism held the business community in awe, the earned compound annual real returns to S&P 500 company shareholders were 5.9 percent. Compare that with 1933 to 1976, a time when business leaders were responsible not simply to shareholders, but to many stakeholders, including employees, local communities, and suppliers; during that stretch of time real returns were 7.6 percent. Ironically, the focus on shareholder return resulted in average shareholders getting fewer returns. As for return on capital, the S&P 500 index is no higher today than it was a decade ago. Financial capitalism based on the efficient market theory hasn't been good for most investors and it hasn't been good for corporations themselves.

Most important, the model has taken a devastating toll on innovation. Because creativity is not considered a key variable, there is insufficient reward for it in big corporations and little pressure to provide incentives for innovation from government. And so most corporations remain locked in a competitive price race, driving profits through efficiency, outsourcing, and cost-cutting while creativity and innovation are shunted to the periphery of the economic system.

IN 2009, MICHAEL MANDEL, A Harvard PhD in economics and former Chief Economist for *BusinessWeek*, presented research that analyzed the failure of the innovation revolution in a cover story called "The Failed Promise of Innovation in the US." Mandel (now Chief Economic Strategist for the Progressive Policy Institute and senior fellow of the Mack Center for Technological Innovation at Wharton) coined the term "innovation shortfall" to describe his shocking conclusion: In the first decade of the twenty-first century, despite all the high-fiving about Silicon Valley and American innovation, most companies weren't innovating at all. If you took away the narrow slices of innovation taking place in communication and looked at the economy as a whole, the picture looked pretty bleak. While futurists in the nineties predicted the United States would switch from a silicon/computer/engineering–based economy to a biogenetic/biology–based economy, Mandel argued advances in biotech science had not yet translated into innovative biotech products. Nanotechnology was supposed to change everything—and didn't. Pharmaceuticals? Vast amounts of spending on research had produced fewer innovative products rather than more.

The consequences of this innovation shortfall have been economically devastating, according to Mandel. With fewer new products to sell, exports (that were not oil or gas) as a share of GDP barely rose from 2000 to 2011. Meanwhile, the accumulated trade deficit in the same period totaled almost $7 trillion, helping fuel the dominance of Wall Street as the principal funnel by which America borrowed from China, Japan, and the rest of the world.

Even in high tech, where it was perceived to be a world leader, the United States fell behind. A 2012 Census Bureau analysis of the imports and exports of ten US high-tech areas, including aerospace, advanced materials, biotech, and life sciences, revealed that a $30 billion trade surplus in 1998 had become a $53 billion deficit by 2007. In 2010, that deficit had grown to $80.9 billion, and by 2011, the trade deficit in high-tech goods had reached $99.6 billion.

Why this amazing decline? A National Science Foundation report released in September 2010, called the 2008 Business R&D and Innovation Survey (BRDIS), showed that only 9 percent of public and private companies engaged in either product or service innovation between 2006 and 2008. This is an extraordinary and unexpected fact. Because of the proliferation of cool gadgets like tablets and smart phones, many of us believe we're living in an era of immense innovation—but we're not. At an Aspen Institute Conference on technology in 2011, Peter Thiel, a cofounder of PayPal and an early investor in Facebook, said that except for big advances in computer-related areas, innovation has actually "stalled out." There's been little innovation in transportation and we haven't seen any new breakthroughs in energy, with alternatives to hydrocarbons making up just a fraction of all energy production.

We have all been feeling the ripple effects of the innovation shortfall. In 1999, the Bureau of Labor Statistics forecast that 2.8 million new jobs would be created over the next decade in "leading edge" industries such as IT, aerospace, telecom, pharmaceuticals, and semiconductor and electronic component manufacturing. Employment in this space would grow at 3.4 percent, twice as fast as the rest of the private economy. It never happened. In fact, leading edge industries *lost* 68,000 jobs over that decade. Even those of us who have jobs have not seen significant enough wage increases. As a consequence of the failure of innovation, Thiel points to the fact that real median wages have, at best, risen by 10 percent since 1973 according to government data. (Other nongovernmental sources show no improvement at all over this time period. All figures are adjusted for inflation.)

Young college grads, who should have been taking advantage of new opportunities in changing industries, actually saw a real decline in wages. Median average starting salaries for college grads fell from $30,000 in 2006 to 2008 to $27,000 in 2009 to 2011 in real terms, according to research at Rutgers University. And while the deep recession certainly played a role, the drop is part of a decades-old decline. The average wage for young college graduates has fallen by nearly a dollar per hour since 2000, accounting for inflation, from $22.75 to $21.77 for men and $19.38 to $18.43 for women, according to the Economic Policy Institute.

To make matters worse, the burden of student loans during this time grew dramatically, so that by 2011, the average college debt for graduating seniors has risen to $23,000, with the total of all student debt topping $1 trillion. As student debt has risen over the past decade, the decline in wages has made it all the more difficult to pay off. In other words, financial capitalism isn't working for the young.

It isn't working for many of the not-so-young either. On September 12, 2012, the Census Bureau came out with a report that showed that median annual household income has fallen back to the 1995 level in real terms. We've experienced not one lost decade of prosperity but nearly two.

Economics has become a discipline that lacks relevance and utility in a world that begs for it. "We need fewer efficient market theorists," says Bradford DeLong, a professor of economics and chair of the political economy major at the University of California at Berkeley. Which presents the question: What do we need instead? What model can replace financial capitalism?

We need to trade in an economics of efficiency for an economics of creativity. Because creativity and uncertainty exist outside the dominant economic model, there is little room for encouraging the role that start-ups play in economic growth. Yet it's start-ups—and larger corporations that haven't lost their connection to their founders—that are, by and large, driving modern innovation and job creation. Why shouldn't our economic model reflect that? We should be moving away from a model of economics based on the "culture of control" that is embedded in the efficient market theory toward a new model that embraces a "culture of chance."

THE RISE OF INDIE CAPITALISM

In 2009, the futurist Paul Saffo predicted that a new "creator economy" would replace the industrial and consumer economies. Writing for management consultant McKinsey & Co.'s online site, Saffo argued that an economy driven by passive consumers was being replaced by one driven by people who created value through their daily activities engaging with companies, providing input and personal data that helped companies develop new products. "The quintessential example of creation is a Google search . . . ," he wrote. "The string of text that goes into the search box seems valueless to the creator, but when aggregated with all the other search strings flowing in, it is valuable enough to make Google worth billions. A simple Google search thus typifies what drives the creator economy—creative value flowing in both directions at the same instant. Other examples of creator transactions abound—think of YouTube and Wikipedia. Interactivity is the common thread, which makes sense because interactivity is what defines the creator economy. . . ."

Saffo's term captures the rise of active consumer participation in one part of the economy, the digital economy. There are, however, deeper and wider changes under way that are transforming the economy. We are becoming makers again, intent not simply on consuming. We are becoming more local, placing more value on national, state, and neighborhood sourcing, design, and manufacturing. We are becoming more entrepreneurial, both online and off.

And so I believe what we're beginning to see evidence of is the birth of a movement I call Indie Capitalism. My use of the word "indie" is deliberate. "Indie" reflects an economy that is independent of the prevailing orthodoxies of economic theory and big business. It shares many of the distributive and social structures of the independent music scene, which shuns big promoters and labels. And as happens with many bands, so many of today's successful creative endeavors began as local phenomena before branching out to new locations and networks. It's no accident that Portland and New York, cities with vibrant indie music scenes, have become hot spots for new businesses that are local, community-focused, organic, socially responsible, and, above all, creative. In my fall 2012 Design, Capitalism and Social Movements class at Parsons, I asked one of my French grad students why she came to New York to study design and creativity if she lived in Paris. She said, "Paris has lost its vibe. I came to New York to find the vibe."

My belief in Indie Capitalism has been inspired by what I've witnessed in companies both new and legendary, and it's bolstered by a single, simple fact: New companies (those less than five years old) have been responsible for all the net new jobs in the United States for the past three decades.

At first, this seems an exaggeration. How could companies so young generate *all* the net new jobs in the United States? But research bears it out; data that the Ewing Marion Kauffman Foundation used from the US Census Bureau in its July 2010 study "The Importance of Start-ups in Job Creation and Job Destruction" showed that between 1977 and 2005 (before the Great Recession and the soaring unemployment rate), existing firms were net job destroyers, losing 1 million jobs net combined every year. In contrast, new firms added an average of 3 million jobs annually. Even during recessions, start-ups tend to remain stable in their employment, while existing corporations fire lots of people. This flies in the face of what we've been led to believe about economic growth. Big corporations are thought to be the drivers of employment and innovation, and our policies and economic incentives reflect that belief.

What would an economy based on innovation and entrepreneurialism, not management and big business, even look like? How would we measure economic growth if we shifted our focus to the development of new value, rather than efficient management of the old?

If Indie Capitalism were to have a single foundational principle, it would be this: Creativity drives capitalism. Indie Capitalism emphasizes the economic value generated by the creation of new products and services. It is an economic system that would encourage the formation of vastly larger numbers of start-ups and promote policies that helped them scale successfully. It would reward large corporations with entrepreneurial cultures—companies like Corning and

3M—that generate substantial revenue each year from new products invented over the previous five years. And it would curtail the "crony capitalism" tactics of big corporations and banks that deploy political contributions to turn the tax and regulatory systems to their benefit.

BUILDING A NEW ECONOMICS OF CREATIVITY

Using the competencies of Creative Intelligence as a starting foundation, we can begin to build a new economic model for the twenty-first century. I developed the central ideas for such a model in the course of teaching Creativity, Capitalism and Social Movement at Parsons with Ben Lee, former provost and Professor of Philosophy and Anthropology at the New School. It is my hope that a community of creators, designers, entrepreneurs, economists, social scientists, and politicians will use these ideas as a jumping-off point for discussions about business practices, policy, regulation, and the economy as a whole.

1. Creativity is the source of economic value.

Creativity transforms what money can't buy into what money can buy. It taps the pools of our aspirations and turns these hopes into products and experiences that have economic value. Profit derives from generating and scaling creativity into marketable commodities.

2. Uncertainty is the state of social and economic life.

Accelerating volatility, complexity, and ambiguity are a constant state of economic life, not an aberration. Uncertainty provides opportunities for innovation, growth, new businesses, and jobs.

3. *The entrepreneur drives economic growth.*

The entrepreneur (including those within corporations who behave like entrepreneurs) is the key player in the model. The entrepreneur, motivated by a calling to create something new or better, generates innovation, growth, jobs, and, ultimately, profits. Established corporations, for their part, innovate best when run by founders or as if the founders were still in control.

4. *Capitalism is a social movement.*

Capitalism is not solely a market phenomenon; it exists within a social context that makes it dynamic, vibrant, uncertain, and ever-changing. Therefore, an understanding of social, political, and technological cultures is central to the model, not an externality. A familiarity with the rituals, rites, behaviors, and values that people find meaningful is key to entrepreneurial activity. Knowing the real and digital cultures of your global networks is a core requirement of managers today.

5. *Social networks are the basic building blocks of the economy.*

People belong to a large and growing number of real and digital communities and play a vast number of roles: consumers, investors, designers, cocreators, makers, thinkers, and collaborators. Significant economic value is generated as a result of the engagements people have within these social networks. Market transactions continue to play an important role, but the social network is paramount in the model.

6. Creative destruction is key to innovation-led economic growth.

An economy based on creativity accelerates the rise and fall of companies. The number and kind of start-ups increase dramatically as they become the dominant actors in the economy. In this model, older, bigger companies fail faster if they cannot change their cultures and innovate.

IF WE USE THESE POLICIES as building blocks for creating a new economics of creativity model, what policies might flow from it? What can we all do, in our lives and our businesses, to promote Indie Capitalism?

Introducing uncertainty into our economic model may appear difficult because it is so unfamiliar to our thinking but, surprisingly enough, the foundation for an economy that centers around innovation and creation might well lie in the ideas of the man whose thinking paved the way for the Chicago School of Economics.

Though he's best known for work on risk and options theory, which helped establish the foundation for financial capitalism and the efficient market theory, Chicago economist Frank Knight also did important research on the role of uncertainty and the entrepreneur in economic growth. In his book *Risk, Uncertainty and Profit*, Knight argues that significant, expanding profits come from entrepreneurs finding new opportunities in the messiness of our lives. "Profit arises out of the inherent, absolute unpredictability of things, out of the sheer, brute fact that the results of human activity cannot be anticipated and then only in so far as even a probability calculation in regard to them is impossible and meaningless."

Knight did not believe in the efficiency of markets, arguing instead that when companies operate for efficiency, and compete on price, profits tend to fall—and sometimes fall to zero. Put another way, efficiency drives a race to the bottom. Creativity, on the other hand, generates products and services that, because of their originality and utility, can have big profit margins and bigger profits.

For Knight, social uncertainty and chance were sources of new knowledge, opportunity, and profits. He was also keenly aware that focusing on "wants" alone was insufficient. "The chief thing which the common-sense individual actually wants is not satisfaction for the wants which he has, but more, and better wants," Knight wrote. Following this line of thinking, opportunity comes from identifying those unspoken "better wants" and creating ways to fulfill them.

But we can't do it alone. If the entrepreneur is to replace the manager and successful start-ups are to replace big business as the drivers of economic growth, then we need to revamp our venture capital system. Right now, we have an archaic way of finding and financing entrepreneurs and their business concepts. A very small number of people—most of them white men, most of them engineers, most of them living on the two coasts—screen potential candidates. While we know of the great successes, it's also commonly known that the overall success rate for start-ups is barely 10 percent. At best, one out of ten projects invested in becomes a successful company.

We need to get the success rate up from 10 percent to 60 or 70 percent and expand the screening of potential creators. Every year, I see dozens of promising senior projects get flushed after graduation. There are literally thousands and thousands of students in colleges and programs around the country and around the world producing ideas and concepts that might make new businesses—but most universities do not have the mechanisms in place to help them do it.

Fortunately, there are "incubators" that provide people with the skills and networks to go from creativity to creation at nearly all the major universities in Silicon Valley. New York City has a growing number and others are opening in Chicago, Cincinnati, and in other cities. But we need thousands of new incubators at all our major schools that can help students take original ideas derived in class and get them out into the marketplace.

Crowdfunding is another means of vastly increasing the number of new businesses, allowing entrepreneurs to circumvent the venture capital hierarchy and giving us all a chance to participate directly in the formation and scaling of start-ups. Kickstarter and Kiva are just a starting point; by tweaking these models, investors could either receive a direct equity stake or lend money and earn a return. As Amy Cortese, a crowdfunding expert and author of the book *Locavesting*, points out, for these sites to become part of the mainstream, "people need to be able to make a financial return. After all, you can't retire on a T-shirt and a film credit!"

Cortese believes that "one of the more successful (and safe) models will be local or community-based crowdfunding." Smallknot, for example, launched in 2012, allows people in a neighborhood to lend money to their local shops. While current securities law doesn't permit them to get interest, they receive free services, gifts, and other "in-kind" payments as a return for helping the shop owners expand or refurbish as well as repay their loans. Egg restaurant in Brooklyn, for example, raised $10,000 for new tables and chairs (all made in New York State) from its own customers. The forty-five investors received rewards in the form of baskets of vegetables and other produce, a biscuit-making class, and a private party for people lending the money.

Smallknot was founded by two young Wall Street finance lawyers, Jay Lee and Ben Rossen, who soured on their experience.

"We were working all hours, moving piles of money from Wall Street to the banks," said Lee in an interview for the *Greenpoint Gazette*, a neighborhood newspaper, in March 2012. At the same time, small businesses couldn't get loans from these banks. They were starving for capital. "We tried to find a way for people to invest money locally," said Rosen.

"Local crowdfunding can mitigate risk," says Cortese. "There is more knowledge of the market and players involved so investors can make more informed decisions. And businesses raising capital will have a built in base of support. Local CF can be a rallying point for communities and form the basis of a local financial ecosystem, where the community capital raised through CF might be matched by banks . . . corporations, local foundations or local government or economic development agencies."

In Britain, crowdfunding models are further along and allow for direct equity investment. Funding Circle, for example, has lent $37 million to six hundred local businesses and Crowdcube has raised about $4 million in equity capital for more than a dozen companies. If Americans shifted just half of their combined $30 trillion of investments in multinational conglomerates to local businesses, argues Cortese, "we'd be living in a far different world."

CREATIVE DESTRUCTION AND CREATIVE EDUCATION

In *Capitalism, Socialism and Democracy*, Joseph Schumpeter wrote, "The fundamental impulse that sets and keeps the capitalist engine in motion comes from the new consumers, goods, the new methods of production or transportation, the new markets, the new forms of industrial organization that capitalist enterprise creates." Schumpeter, the father of "creative destruction," saw a version of capitalism in which entrepreneurs were the central, disruptive energy sustaining economic growth. The process of creative destruction undermined established monopolies and toppled existing big businesses that made profits from older regulatory, organizational, technological, and ideological regimes.

We see Schumpeter's creative destruction thriving in Silicon Valley and among start-ups in general. Apple is pushing aside RIM and Nokia, Facebook is pressuring Google to move into social, online shopping is mauling the malls. But the brutality of the free market isn't in operation in many industries where large corporations hold sway by virtue of their powerful connections, not their innovations. Creative destruction is the enemy of crony capitalism, as well as a force that's essential for innovation.

An economic model based on creativity would require a transparent economic playing field, a vigorous antitrust policy that breaks up unfair monopolies, and curbs lobbyists who manipulate tax and regulatory policies for special interests. The capital gains tax, for example, should be structured to promote start-ups and young companies. And a trade policy must reflect the needs of new companies and the "local" values of a growing number of people.

On the individual level, an economy based on creativity rather than efficiency naturally requires a different set of skills. It's never too late to learn these skills, but we should begin nurturing them early on by moving our educational curriculum away from "teaching to the test" and toward the idea of "learning for creativity." An emphasis on science, math, and engineering will help students meet tomorrow's technological challenges, but the way these subjects are taught is every bit as important as the information shared. Students need to learn more than just numbers and formulas; they need to understand the context of this information they're being taught, not simply the "right answers." And they need room to play.

We must also remember that creativity emerges out of a social and cultural context. Leaning that context, as well as the skills of creativity to operate within that context, will require the foundation of an expanded liberal arts education that includes the literacies of writing, reading, and art. If writing computer code is now considered an additional form of composition, then art, music, acting, and dance must also be seen as key modes of expression—competencies that may well inspire a budding entrepreneur just as calligraphy inspired Steve Jobs.

We should go further and reframe our notion of a liberal arts education to design a new "creative arts" education. Embedding the creative competencies of Knowledge Mining, Framing, Playing, Making, and Pivoting where they are age-appropriate is a first step in constructing a pedagogy of creativity that we should all learn from kindergarten through university.

THE CONTOURS OF INDIE CAPITALISM are already in sight at the edges of our economy. If we master the right competencies at the individual level and employ the right policies in business and government, Indie Capitalism would take the following shape.

- Indie Capitalism would be more local and less global. Making things locally and in the United States would be a key priority, and this emphasis on localization would also change the way many global corporations operate domestically. The shift is beginning at a number of companies, with Boeing, Caterpillar, and GE "reshoring" a small part of their production back to the United States and publicizing the move in high-profile ad campaigns that reflect a growing demand for more domestic production. Finally, localism will require global corporations to adopt the values of the American local movement, paying workers higher wages, improving working conditions. It will mean pressuring American corporations to take pride in the label by paying a higher percentage of taxes rather than "booking" them outside the country specifically to avoid the IRS.

- Markets will continue to play a critical role in Indie Capitalism, but the blurring of distinction between creators, curators, funders, and consumers will make it less transactional. People will participate more directly in every step of creation, using social media and making their own choices about the design, financing, and consuming of the products, services, and experiences they want. They will extend their "playlist skills" beyond their music collections and to the entire spectrum of business, health care, education, and eventually politics.

- Indie Capitalism would provide incentives for people to become craftsmen and craftswomen and not simply passive consumers. The ability to put down a clean weld would become a highly valued skill, as would sewing the perfect stitch. Mastering tools and making great things would become a routine part of a meaningful existence, replacing consumption as an end in itself.

- Just as the industrial economy required an education system that prepared people for work in factories, Indie Capitalism will require a system of learning that equips people with the competencies to create. We need an education system that teaches people how to

connect very different domains of knowledge. Children and adults should be taught how to frame and
reframe their interactions and understand how narratives convey meaning. They should be taught how
to puzzle out, rather than problem solve, and that in
many situations, there are many right solutions. They
should learn that playing and discovering are often
better techniques than trying to get it right.

Despite promising movement toward an alternative economic model from start-ups and a few global corporations, there
is still much work to be done to overcome widespread belief in
old and stale models and ideologies. Whether you are from the
United States, Europe, India, China, Brazil, or somewhere else,
you are still likely to find yourself having that conversation about
big government vs. small government or whether capitalism is
good or bad. As we're indulging in the same tired conversations,
centripetal forces are pulling our societies apart, leaving us adrift,
wondering how to shape our lives in an ever-changing world.

In the heat of debate, we cannot hear what we actually agree
upon. In our anger and frustration, we cannot see the solutions
we can commonly support. An economics of creativity model that
promotes a new form of capitalism, Indie Capitalism, is something that breaches the silos of "Democratic" and "Republican."

Those working in professions that once seemed noble, even
heroic, no longer command our respect. Politicians and business leaders have fallen in popularity everywhere, within all age
groups and especially among the young. But a few heroes remain,
and the entrepreneur is one of them. The death of Steve Jobs was
mourned by Republicans and Democrats, old and young, Chinese
and American.

And we still have at least one enemy in common. Call it financial capitalism, shareholder capitalism or crony capitalism; this outmoded convention draws the ire of both left and right, 99 percenter and Tea Party member alike.

We celebrate the entrepreneur because we value the entrepreneur's creativity. It is that creativity that we need to make central to our economy and to our economic thinking. The data clearly show that the efficient market theory has fostered an economic system that, over the past two decades, has generated little innovation among most companies, weakened the middle class, widened inequality, and led to the relative decline of the United States.

This has taken a terrible toll on all of us who seek to make our mark, establish a career, and build a life in a very difficult era. But by learning the competencies of Creative Intelligence, by building an economics of creativity model, and by shaping a new Indie Capitalism to foster growth and jobs, we can innovate a much stronger future.

The need to reconnect creativity to capitalism has never been greater.

What's Your CQ?

SIXTEEN OF US HAD SPENT two days talking about innovation and design thinking at the Stanford d.school in March 2009 when the Executive Director of Stanford's design program, Bill Burnett, framed one of the great challenges of our day. It had been the third Summit on the Future of Design and a high-powered group of educators and practitioners who specialized in innovation and creativity had pondered the question: What will define the next frontier for innovation and what new methodologies would need to be developed?

The conference focused on what large corporations needed to do to become more innovative, and the discussion revolved around how we might go beyond design thinking to something richer, something that reflected the intelligence necessary for breakthrough design that was deeper than the process of thinking alone. The ideas and discussions I had over that weekend pushed my own thinking about the social and cultural aspect of creativity and helped me formulate some of my ideas for assessing creativity.

It was at the end of the second day that Burnett, who had worked at Hasbro and Apple, made a statement that really resonated with me. "We have GREs and SATs to measure math, verbal, and writing scores," he said, "but we don't even attempt to measure creativity. We have no currency with people who do measure." By that Burnett was describing admissions officers at schools like Stanford, but he could have been talking about HR departments in corporations and other organizations. Burnett said that whether we like it or not, "What can't be measured doesn't have value." And then he concluded with this: "We need a measure of creativity. We need an SAT for creativity."

My close friends have a four-year-old daughter named Zoe, and my hope is that by the time she applies to Stanford (and she will), the university admissions office will have developed a methodology to assess her Creative Intelligence. Perhaps by then we will have comparable "CQ assessments" for children around the world, paralleling the current global measures on math and science. How would American students match up against German or Korean students in creativity? How would they fare against French or Brazilian kids? Would students from New York and London, San Francisco and Milan, where there are many schools in design, fashion, art, media, music, and architecture, do better than students from places without these kinds of educational opportunities?

How should we go about building a method to assess creativity on a national and global scale? Where do we start? As I discovered from my investigation into the history of creativity research, the attempt to find quantitative measures of creativity hasn't been very fruitful in the past. But what's the alternative? We do, after all, tend to associate assessment with mathematical measurement. From the incessant testing that goes on in K–12 classes thanks to No Child Left Behind, to Six Sigma management in business, we constantly test by the numbers.

But creativity doesn't appear to lend itself to metric measurement at this point in time. Perhaps in the future, we might be able to come up with an algorithm that works. But for now, we need qualitative measures. So I set about talking to the most creative people I know about what to do, asking them, "How do we assess creativity?"

I first turned to what might appear an unlikely source, a business school dean. The overwhelming number of these deans were dedicated evangelists of the efficient market theory model and the purely mathematical analytical skills that go with it. Nothing creative about the concept or the competencies taught in their schools. But Roger Martin at the Rotman School of Management was among the first deans to bring design thinking into his school's business curriculum.

Martin offered up some simple and obvious advice: "Do what Juilliard does," he said. "Look at students' portfolios. And test them on their performance." In other words, go to places that already assess creativity. It was good advice. The Juilliard School's process for auditioning talented dance students begins with prospective students attending ballet classes where they are observed by a teacher. The basic level of training and ability is judged, and some receive callbacks to perform modern dance. In effect, their "portfolios"—what they've learned and how they move—are assessed.

Then each student performs a two-minute solo dance of his or her choosing, either original choreography or from repertory. Another assessment is made, and a small number are called back. At this point, the student is briefly coached on modern dance movements and told to perform right there before a panel of faculty members. The panel assesses how fast the student picks up choreography, responds to corrections and changes, and works as a member of an ensemble.

The Juilliard process includes both planned and unplanned or spontaneous performances. The professors look at technique as well as risk-taking, skill as well as learning ability, solo performance as well as ensemble work. All these categories are assessed by a small jury of experts who, based on their own domain knowledge, determine the best candidates.

This method of assessing Creative Intelligence can be found

throughout society. Creativity is assessed using portfolio and performance criteria every day, Olympics gymnastics and on music sharing sites.

Nowhere is this method of assessing creativity more evident than on TV. Turn your TV on and you, along with tens of millions of other people, can see shows like *Project Runway* or *Chopped* whose judges assess skill and originality much like Juilliard does. Most competition-style reality TV shows use a similar model: small jury, specific challenges, performances, assessments, scores, and winners. On *Dancing with the Stars*, amateurs are paired up with professional dancers, although the "amateurs" are often celebrities in fields other than dancing. Astronauts, Olympic athletes, singers, and supermodels—people with high visibility—are brought onstage with the pros. A three-person panel of judges consisting of professional dancers then scores them on points. The audience also votes. Couples that receive the lowest combined scores are eliminated every week until only one pair remains dancing. In this model, you have a panel of experts combined with a large group of nonexperts providing an evaluation of the performance.

Feedback from "expert" judges is important. As discussed earlier in the book, researchers have discovered that creativity emerges from specific fields of knowledge. Experts—whether it's a fashion designer on *Project Runway* or a celebrity restaurateur on *Top Chef*—share an ability to judge what is different and what is not, what is truly innovative and what is not.

Mihaly Csikszentmihalyi was one of the first to use a jury of experts to evaluate creativity. In *The Creative Vision*, which Csikszentmihalyi wrote with Jacob W. Getzels in 1976, he describes telling MFA students to paint still-life pictures and asking a panel of five art experts to rate each of them on originality, craftsmanship, and aesthetics. Teresa M. Amabile later used a jury to assess creativity in schoolchildren by looking at their collages. This technique of using a small jury of experts to assess performances or a body of work has been codified into a methodology called the consensual assessment technique (CAT) by Amabile and colleagues.

I wanted to get into the business world and see who was using an assessment model to find creative employees. So I talked to one of the most creative people I know, Tim Brown, who runs one of the most innovative consultancies around, IDEO. In 2012, 7 percent of all business school graduates said they wanted to work at IDEO. That placed the consultancy thirteenth on a list of the one hundred favorite companies to work for, according to research firm Universum.

On March 28, 2011, Brown presented in my Design at the Edge class at Parsons. It was a tour de force about design and innovation, and the eighty-seven students loved it. In speaking about the need for change, Brown said that businesses and schools needed to rethink the whole notion of a résumé. It's a bad measure of what they are really seeking. "The résumé is a nineteenth-century idea. It's ridiculous," he said. So how do they hire?

IDEO begins with what people have already done, assessing applicants' portfolios. "The portfolio is the architected communication of what it means to be you and what you've done in the world," said Brown. It extracts your experience and abilities and presents them in tangible form to others. A portfolio should be meaningful to the expert who is assessing it and provide that person with a way of calculating your future capabilities. "That's a far better way of understanding somebody than a résumé," Brown said as part of his presentation.

At IDEO, portfolios are presented in several forms. They can be movies, interactive games or sites, visual imaging, or even just the written word. "I don't mind how conceptual it is as long as it is tangible," said Brown. For members of Gen Y who grew up using Apple's digital tools, making YouTube videos, generating images for Instagram, and building their own Facebook pages, portfolio works particularly well.

But how do companies find people who may be creative but don't see themselves that way (most of us, really)? What about the business and science majors who don't have a formal portfolio to show? IDEO may have an assessment technique for that too. Ideo.org, a nonprofit organization that came out of IDEO's work in the social sector, asks people to build a "visualization of self," a revealing portfolio of their hobbies, interests, travels, aspirations. No matter your field, constructing a portfolio—virtual or physical, highly designed or not—is a way of holding a mirror to your creative accomplishments. Not only will such a portfolio make you more attractive to potential employees than a standard résumé; it's a way of making yourself aware of creative skills.

For IDEO, portfolio assessment is just the beginning of the vetting process. Then comes the performance. IDEO, like Google and a growing number of companies, gives performance challenges right at the interview. And they give up to six interviews per prospective employee. Most often, people are asked to perform in teams.

A lot of creative companies outside the realm of performing arts have similar hiring processes. They don't just hire you and hope for the best; they give you assignments, then assess you on your performance. Spotify, for example, has a page of puzzles that assess problem-solving abilities, which it encourages prospective employees to try. They are mostly complex math puzzles that only the best writers of software code can solve.

Continuum, the innovation consultancy that designed the new on-the-go Tetra Pak DreamCap, also combines portfolio with performance when evaluating the creative capacity of applicants. It gives a test on the spot to see if people have the raw creative capabilities necessary for the job. "These are especially important for people who might not have a conventional design background or portfolio," said Harry West. One of them is the bottle test. A bunch of cans, juice bottles, and packs and other liquid containers are brought into a room and the candidates are asked to describe them. Can they see the differences in color, shape, labeling? Can they see their relationship to the content, the drink itself? And most important, can the candidate frame a narrative story about the drink based on the design elements—who is it for, what is it really, how will it be drunk? "Some people smile and dive deep, creating interesting stories to connect the cues in the packages in front of them," says West. Those are the people who get invited back. At the same time, "Some people are deer in the headlights: They can't see design," says West. "They wonder why they are being asked to do this." Those candidates fail the audition and don't get a callback.

In judging portfolios and performances, Continuum has a clear goal in mind, one that differs from conventional design thinking and innovation strategies. "We are not interested in random ideas," says West. "We are not interested in a hundred ideas, but we are interested in the right idea which can align a complete solution." That means people who can bring many different aspects of an issue together with a single concept.

WHAT CAN YOU, AS AN individual, do to boost your Creative Intelligence? A good first step is to stop and simply reflect about what you are good at, what you can perform. Most of us don't know how to evaluate our own creative competencies. We often are not aware of our own capabilities, and when we are, we fail to perceive them in a larger context. We're not recognizing how to connect the skills from one area to another. We're not seeing them as skills or presenting them as skills. Many of us lead book clubs or sports teams but don't recognize that we're actually creating "magic circles." You might be great at reading body language or bargaining while shopping or organizing family trips but don't realize that by framing these skills differently, you can utilize them in countless creative ways.

One way to make yourself aware of your creative potential is to start keeping a portfolio. A portfolio could begin as a journal that contains your ideas, notes, sketches, and work. Some people could include fashion concepts. Others, new business models for start-ups. A portfolio could show you the many "dots" you've collected over the course of your life, make them transparent and alive, and encourage you to begin connecting some of them.

Colleges should begin to request portfolios as part of their admissions process—and not just design schools but business schools as well. Corporations that are not yet doing so should be requesting portfolios as part of their hiring process and adding performance challenges to their job interviews.

There may already be a model for assessing the CQ of hundreds and perhaps thousands of people at a time—a game called Odyssey of the Mind. Millions of students from kindergarten through college have played it. There are tournaments in cities and towns all over the world leading up to the world finals every year. Students play in every state of the United States, Britain, Korea, Mexico, Poland, Germany, China, Uzbekistan, Japan, Russia, Argentina, Kazakhstan, and other countries.

The game was created thirty-four years ago by Sam Micklus and Theodore Gourley at Rowan University of New Jersey (then Glassboro State College). It had its beginnings in the classroom of Micklus, a professor of industrial design who began to challenge his students to create cars without wheels, mechanical pie throwers, and other fun stuff that he then judged on ingenuity and risk-taking, not success. Word got out that this was fun. It grew into a larger competition and eventually into Odyssey.

The goal of the game is straightforward. According to its website, "The Odyssey of the Mind teaches students to learn creative problem-solving methods while having fun in the process. Students learn how to identify challenges and to think creatively to solve those problems."

In the game, there are age divisions, from grades K–5 (less than twelve years of age) through college, and all but the youngest compete. Every year, students in teams of up to seven are offered five Long Term Problems that range from the technical to the artistic. The problems fall into five general categories: Mechanical/ Vehicle, where teams build and operate actual vehicles; Classics, where the problem is drawn from literature or art; Performance, which involves stage performances around themes; Structure, where teams build houses and other structures; and Technical Performance, where teams build robots or gizmos that incorporate artistic elements. In the 2011 to 2012 game, one challenge was to put on a musical theater performance of Shakespeare's line "To be or not to be." A previous year's problem was to design and run vehicles whose only source of energy was mousetraps.

The teams are then assessed by small panels of three to six judges. There are two areas of competition, the Long Term Problem and something called Spontaneous. As at the Juilliard audition or the IDEO interview, students are given a problem on the spot by the judges to solve in a creative way. The winners of all the tournaments and of the world final have to come out best on both the Long Term Problem and the Spontaneous Problem. It's similar to being assessed on Portfolio and Performance.

Judges are mostly parents and other adults who volunteer to give up some serious personal time. Problem Captains—experts who have years of experience observing student projects—organize and train the judging teams. The game has a Head Judge, who leads a specific judging team that compiles the assessments made on score sheets; a Staging Judge, who checks teams in at their competition time and makes sure everything is in order; and other types of judges. For a single statewide tournament, up to seventy judges—all volunteers—can be needed.

Odyssey of the Mind is a good existing methodology for assessing creativity on a national and even global basis—perhaps not as good as the more formal and better-financed Olympics, but pretty close. Both show that scoring portfolios and performance is an excellent way to assess creativity. It's not an "SAT of creativity," but definitely a step in the right direction.

Kickstarter is yet another model of assessing creativity on a large scale. It combines both the "expert" and the "mass" models of assessment. A small group of curators (it used to be just the three founders and has now expanded to many more to keep up with the explosion of proposals) screens all the applications for new books, movies, fashion lines, games, dance performances, watches, food, and other projects. These "judges" use their expertise to select a limited number of candidates, which are posted online. Then the crowd assesses them, voting with their patronage. This hybrid model of assessing creativity is a tremendous success. It's a different kind of judge and jury but a similar kind of assessment.

Kickstarter is also a wonderful window into the way thousands of people are being creative in thousands of different ways. Check it out and see the amazing array of products and services being proposed that you probably have the potential to create.

The challenge ahead of us is not to invent new forms of assessment for creativity. We've already done the hard work. We've figured out that portfolios and performances can provide a pretty good picture of our creative competencies. And we've learned that small groups of judges can validly determine what is creative and what is not in different content areas. This is an important and rather amazing accomplishment that should not be taken for granted.

What we need to do now is apply the models that are all around us—on our TV programs, in our multiplayer games, and in our crowdfunding ventures—to our own personal lives and businesses. We need to move the way we assess creativity from the periphery of our lives to the center of our society.

Rethinking Creativity

IN AUGUST 2012, APPLE BECAME the most valuable company in history. Its soaring stock, which hit $680 per share, pushed the company's market capitalization toward $700 billion. Just a year before, ExxonMobil had been at the top of the pyramid. Now it lagged far behind as global sales of iPhone and iPads skyrocketed. Of course, Microsoft had been number one back in 1999 at the top of the Internet boom. And General Motors and IBM had once been dominant in their day as well. But unlike that of previous corporate achievers, Apple's value did not rest on energy extraction, manufacturing, or even technology. Apple is the first company to become "most valuable" because of its creativity.

In pricing Apple so highly, the market is beginning to value creativity in a new way. It is recognizing, perhaps for the first time, that the greatest economic value can now be found in the kind of originality that engages us, satisfies our aspirations, and empowers us to make our lives better. In valuing Apple so highly, the market is sending a clear message that building new things that possess a special aura, an attraction that pulls us toward them, can be more important than merely making (or copying) things to sell at a lower price.

We need to heed this message. On a policy level, the increasing value of creativity bolsters the idea of an economic system based on innovation. It makes even clearer that Washington should be focused on promoting creative skills in the classroom and entrepreneurial ventures in the business world. This is a very different agenda from the one we currently have.

There are clear lessons as well for policy makers the world over, and especially in Asia. Asian business leaders have proved to be innovative in the past—Sony's Walkman began the era of mobile music, after all—but the thrust of government economic policy in Korea, China, and Taiwan for decades has been to support "fast followers" of Western innovation.

That may be beginning to change. Korea has spent tens of billions of dollars on design to reach parity with Europe and America in its phones, TVs, and cars. Thousands of Korean students are attending design schools in the United States and Europe. The rising economic value of creativity now means Samsung, LG, and other Korean corporations must take the next step and break through the barrier of originality to truly innovate.

China, too, is trying to change course. The "faster, cheaper" economic strategy that lifted tens of millions out of poverty was built on importing and absorbing innovations at a low cost. That era appears to be ending as the cost of both labor and importing Western innovation rises sharply. Recognizing this, China's government is already changing economic strategy. In announcing its latest Five Year Plan in 2011, China's National People's Congress called for higher-quality growth in aerospace, biotech, and other new industries through "indigenous innovation." How to promote that local creativity within an authoritarian social and political system will be China's, indeed much of Asia's, challenge ahead. Creativity and freedom are mutually reinforcing. Creativity and authoritarianism are not.

The rising value of creativity has important implications for each and every one of us. Just as Apple's market value is rising, so too is the value of our own creative competencies. Creative people who can go deeper, frame new patterns and engagements, and use their knowledge to build wildly engaging products should now be worth more in the marketplace. Degrees from colleges and universities that teach creative competencies and business schools that show people how to be more entrepreneurial should gain in value as well.

But first we must recognize the value of creative competencies and a creativity-driven society, and unfortunately the current level of political discourse will make it very difficult to do so. Despite our vast creative potential, we are everywhere—the United States, Europe, Asia—mired in ways of thinking that are archaic, rigid, and insufficient. We talk and think in ideologies and categories that are no longer relevant.

We need to stop arguing about big government vs. small government, about which side is right and which is wrong. Balanced budgets, taxes, and regulations are all important issues, but discussions revolving around conflicts have a way of devolving into "either-or" arguments. We're stuck in a problem-solving mindset—as if there's one correct solution to any of the issues facing the nation, as if the puzzle will end as soon as we get it right.

We need a twenty-first-century reframing of ourselves and our institutions. We must view the world and our economic and social challenges not as an argument, but as a game: one that, hopefully, has no end, but simply requires continual learning and discovery. We must reframe our problems as an exploration and in so doing perhaps rediscover what unites us.

Most of us know that the economic and political superstructures of our lives are irrelevant to us. We should take a moment and simply recognize this truth and focus on the things we as individuals, and organizations, can do. And there is so much we can do, so much we can create.

If there is one great challenge ahead, it is not about the size

CREATIVE INTELLIGENCE / 266

of the budget deficit or the role of markets in society. The biggest challenge we face is our own fear of creativity. Our social tensions and economic failures can be met only by exploring new avenues of possibility. All the great challenges of our day are connected to a need for us all to recognize our creativity and hone our creative abilities so we can find those pathways of possibility. Yet we have been brought up to believe that creativity is rare, the special gift of a few individual geniuses, the magical quality that we don't have and can't share. This destructive myth of creativity has crippled us as individuals and as a nation.

Every time I see little four-year-old Zoe play at turning a basket into a hat, paint the most brilliantly colored picture, or spin a completely original story about herself piloting a plane across the ocean, I know how untrue that myth is. We can all be creative. We just need to get back into practice.

And in practicing creativity, in using our Creative Intelligence, we can build better careers for ourselves, new kinds of businesses, and health and education systems that make sense in the twenty-first century. We can reinvent and revitalize our capitalist economy and take it to the next level. It's a liberating and exciting prospect and maybe, at times, even risky. But what else are we here for if not to create a better world than the one we inherited?

Notes

CHAPTER 1

3   In 1959, a young man: Keith Richards and James Fox, *Life* (New York: Little, Brown and Company, 2010), 65.

3   he'd hated school: Ibid., 28.

3   "adopted a criminal mind," Ibid., 56, 65.

4   "At a pace that seemed wholly un-British": Shawn Levy, *Ready, Steady, Go!* (New York: Doubleday, 2002), 4, 5.

4   "You realize later on that": Richards and Fox, *Life*, 50.

4   His teenage years were not: Ibid., 42–48, 56–57, 65.

4   Even at Sidcup: Ibid., 67–69, 78–81, 84.

5   In 1962, the boy: 87–90, 99–100.

5   the Stones have grossed: Andy Serwer with Julia Boorstin and Ann Harrington, "Inside the Rolling Stones Inc.," *CNN Money*, September 30, 2002, accessed September 4, 2012, http://money.cnn .com/magazines/fortune/fortune_archive/2002/09/30/329302/ index.htm.

6   I've heard of an engineer: personal record.

6   I've seen a student: personal record.

7   Researchers at Cornell University: J. S. Mueller, S. Melwani, and J. A. Goncalo, "The Bias Against Creativity: Why People Desire but Reject Creative Ideas" (electronic version), retrieved July 2012 from Cornell University, ILR School site: digitalcommons .ilr.cornell.edu/articles/450/.

7   And there have been a number of studies: Kay Redfield Jamison, *Touched with Fire: Manic Depressive Illness and the Artistic Tem-*

perament (New York: Simon & Schuster, 1993); Kate Stone Lombardi, "Exploring Artistic Creativity and Its Link to Madness," *New York Times*, April 27, 1997, accessed October, 15, 2012, http://www.nytimes.com/1997/04/27/nyregion/exploring-artistic-creativity-and-its-link-to-madness.html.

7   Einstein's brain was extracted: "The Long Strange Journey of Einstein's Brain," NPR broadcast, http://www.npr.org/templates/story/story.php?storyId=4602913, April 15, 2005; Brian Burrell, *Postcards from the Brain Museum: The Improbable Search for Meaning in the Matter of Famous Minds* (New York: Broadway, January 11, 2005).

7   (Lenin's brain was also examined this way): Andrew Higgins, "Lenin's Brain," *Independent*, November 1, 1993, accessed September 4, 2012, http://www.independent.co.uk/life-style/lenins-lenins-brain-they-took-it-out-to-understand-the-source-of-a-revolution-they-now-reject-but-they-tend-it-still-safe-and-sound-in-31000-pieces-andrew-higgins-reports-1501441.html; Paul R. Gregory, *Lenin's Brain and Other Tales from the Secret Soviet Archives* (California: Hoover Institution Press, 2008), 25–26.

8   He played violin, originally: Brian Foster, "Einstein and His Love of Music," *PhysicsWorld*, January 2005, accessed September 13, 2012, http://www.pha.jhu.edu/einstein/stuff/einstein&music.pdf.

8   He struggled in school: Barbara Wolff and Hananya Goodman, "The Legend of the Dull-Witted Child Who Grew Up to Be a Genius," accessed September 13, 2012, http://www.albert-einstein.org/article_handicap.html.

8   We also know Einstein drew: John S. Rigden, *Einstein 1905: The Standard of Greatness.* (Boston: Harvard University Press, 2006); Albert Einstein; John Stachel, ed.: *Einstein's Miraculous Year: Five Papers That Changed the Face of Physics* (Princeton, NJ: Princeton University Press, 2005).

8   While Einstein was a patent clerk: "Einstein in the World Wide Web: Akademie Olympia," accessed September 13, 2012. http://www.einstein-website.de/z_biography/olympia-e.html.

8   Einstein acknowledged the effect: Carl Seelig, ed.; Sonja Bargmann, tr.; *Albert Einstein: Ideas and Opinions* (New York: Crown Publishers, Inc., 1954).

8   And we know that Keith Richards: Richards and Fox, *Life*, 142–43.

8   Oldham, who'd worked for Mary Quant: Ibid., 127–30; Andrew Loog Oldham with Simon Dudfield and Ron Ross, ed., *Stoned: A Memoir of London in the 1960s* (New York: St. Martin's Press, 2000).

8   It was because of their collaboration: Richards and Fox, *Life*, 142–43.

9   "What I found about the blues": Ibid., 94.

CHAPTER 2

11   In 1992 *BusinessWeek* began supporting: Bruce Nussbaum, "The Best Product Designs of the Year," *BusinessWeek*, June 7, 1992, http://www.businessweek.com/stories/1992-06-07/winners-the-best-product-designs-of-the-year.

11   What is now called the International: The IDEA was originally called the Industrial Design Excellence Awards and changed in the late nineties to be more global.

11   In April 2006, the magazine teamed up with: Bruce Nussbaum, "The World's Most Innovative Companies," *BusinessWeek*, April 23, 2006, www.businessweek.com/stories/2006-04-23/the-worlds-most-innovative-companies.

11   Then, in 2008, my team: Bruce Nussbaum, "S&P/BusinessWeek Global Innovation Index," *BusinessWeek*, May 1, 2008, www.businessweek.com/stories/2008-05-01/s-and-p-businessweek-global-innovation-indexbusinessweek-business-news-stock-market-and-financial-advice.

11   Between our many lists, the index: Bruce Nussbaum, "*BusinessWeek* Is Launching INside Innovation," *BusinessWeek*, May 13, 2006, http://www.businessweek.com/stories/2006-05-13/business-week-is-launching-inside-innovation-dot; http://www.designing-media.com/interviews/BruceNussbaum, accessed October 15, 2012.

13   Frustrated, I began asking around: personal record.

13   from Kaiser Permanente's shift: Bruce Nussbaum, "The Power of Design," *BusinessWeek*, May 17, 2004.

13   to P&G's development of the Swiffer: Bruce Nussbaum, "Get

Creative! How to Build Innovative Companies," *BusinessWeek*, August 1, 2005.

13   Years after I initially began these conversations: personal record.

14   I had joined Danny Hillis: I'd been covering design for more than a decade by the time of this meeting and was on a first-name basis with most of those attending it. They were what management consultants called "thought leaders" in innovation and design. They were, in fact, among the leading thought leaders of the field in the world. I've spoken at conferences organized by Patrick Whitney and have been onstage with Roger Martin and Larry Keeley.

15   While most Fortune 500s were no longer: Davis Dyer, Frederick Dalzell, and Rowena Olegario, *Rising Tide: Lessons from 165 Years of Brand Building at Procter & Gamble* (Boston: Harvard Business Review Press, 2004).

15   With products like the Swiffer mop: Henry W. Chesbrough, MIT Sloan Management Review, "Opening Up Innovation at P&G," posted on InnovationTools (January 18, 2008), accessed September 13, 2012, http://www.innovationtools.com/Articles/EnterpriseDetails.asp?a=297; Ben Thompson, "Designer Thinking," *Business Management*, issue 12, accessed September 13, 2012, http://www.busmanagement.com/article/Designer-Thinking/; A. G. Lafley, "P&G's Innovation Culture," with an introduction by Ram Charan, strategy+business.com, http://www.strategy-business.com/article/08304?pg=all.

15   Over at the Stanford "d.school": David Kelley interview with author at Stanford D-School (August 9, 2009).

15   Though a longtime champion: I first interviewed David Kelley even before he merged his company, David Kelley Design, with ID Two, founded by Bill Moggridge, and Matrix Product Design, Mike Nuttall's firm, to form IDEO. That was in 1991. While writing a cover story on IDEO in 2004, I rode around in one of Kelley's vintage trucks and visited his Ettore Sottsass–designed house. His focus on human-centered design has always been an inspiration to me.

16   In 2010, IBM ran a survey of 1,500 CEOs: Adam Richardson, "IBM Study: CEOs Say Creativity and Managing Complexity Are Vital Today," posted on CNET News (May 26, 2010), http://

news.cnet.com/8301-13641_3-10474433-44.html; "What Chief Executives Really Want," *BusinessWeek*, May 18, 2010, accessed September 13, 2012, http://www.businessweek.com/innovate/content/may2010/id20100517_190221.htm.

16   I was fortunate to find out that I wasn't alone: Among the others at the summit were Nick Leon, now director of Design London; Bill Burnett, Executive Director of the Stanford Design Program; Ronald Jones, who leads the Experience Design Group at Konstfack University College of Arts, Crafts and Design in Stockholm; Nathan Shedroff, who was then launching a design strategy MBA at California College of the Arts in San Francisco; and Jacob Mathew, cofounder of IDIOM, the top innovation consultancy in India.

17   Who coined those phrases: I've asked several of the participants at the Stanford conference who they thought came up with the terminology first, and Nick Leon, Bill Burnett, and Banny Banerjee are most often mentioned.

17   On a cold winter night: William Casey, personal conversation with the author. Off the record, unless you count what the CIA recorded without the author's knowledge or consent. Just kidding.

17   That book, *The World After Oil*: Bruce Nussbaum: *The World After Oil: The Shifting Axis of Power and Wealth* (New York: Simon and Schuster, 1983).

17   Casey had been head of the CIA: Joseph E. Persico, *Casey: From the OSS to the CIA* (New York: Viking Penguin, 1990), 53–73.

17   But Casey's career in espionage: Ibid., 53–67.

18   Donovan immediately liked Casey: Ibid., 55.

18   Despite his talents: Ibid., 56.

18   Determined to prove himself: Ibid., 58.

18   In December of 1944, the thirty-one-year old: Ibid., 68–70.

18   Casey's SI agents were given: Ibid., 71–73;

19   Casey told me, in his thick: personal interview, William Casey with author, winter 1983.

19   According to one report: Report of Proceedings, Joint Special Operations University (JSOU) and Office of Strategic Services (OSS) Society Symposium, "Irregular Warfare and the OSS Model," (Tampa, Florida: JSOU Press, 2009), 19.

19   OSS trainees were encouraged to: John Whiteclay Chambers

II, "Training for War and Espionage," from chapter "Office of Strategic Services During World War II," *Studies in Intelligence*, vol. 54, no. 2 (June 2010), 17.

19   I asked him what happened to the people: My conversation with Bill Casey happened decades ago. But while I may not remember his or my own exact words—I didn't take notes—this was my one and only talk with the head of the CIA and the scene and discussion are clear in my mind.

19   Moe Berg, a Columbia Law School grad: "The Catcher, the Physicist, and the Nazis' Bomb," D.Chieftan.com, http://www.dchieftain.com/2010/10/13/the-catcher-the-physicist-and-the-nazis%E2%80%99-bomb, accessed September 27, 2012.

20   The actor Sterling Hayden: Richard Harris Smith, *OSS: The Secret History of America's First Central Intelligence Agency* (Berkeley: University of California Press, 1972), 133.

20   In his book *Explaining Creativity*: R. Keith Sawyer, *Explaining Creativity* (New York: Oxford University Press, 2012). Sawyer is one of the great chroniclers of modern creativity research, and his work and writings were critical to my own understanding of creativity. I came to the subject from a different stream of knowledge, design, and innovation, and Sawyer provided the historical context necessary for me.

20   One of these men was J. P. Guilford: Ibid., 15–16.

20   During the Cold War: Ibid., 37.

20   The NSF funded a series of key: Frank Barron and David M. Harrington, "Creativity, Intelligence, and Personality," *Annual Review of Psychology*, vol. 32 (1981): 439–76; Sawyer, *Explaining Creativity*, 17.

20   By the sixties: Sawyer, *Explaining Creativity*, 17.

21   In 1960, Ellis Paul Torrance: Ibid., 17.

21   But despite the widespread use: Ibid., 50–53.

21   In the eighties, Teresa M. Amabile: Ibid., 83; T. M. Amabile, *The Social Psychology of Creativity* (New York: Springer-Verlag, 1983).

21   As Sawyer points out, Amabile concluded: Sawyer, *Explaining Creativity*, 60; Amabile, "The Social Psychology of Creativity: A Consensual Assessment Technique," *Journal of Personality and Social Psychology*, 1982, vol. 43, no. 5, 997–1013.

21   Amabile's research also marked: "Faculty Profile Page: Teresa M.

Amabile, Edsel Bryant Ford Professor of Business Administration Director of Research Harvard Business School," accessed September 5, 2012, http://drfd.hbs.edu/fit/public/facultyInfo .do?facInfo=bio&facId=6409.

22   By the time Amabile and others: Sawyer, *Explaining Creativity*, 52–56.

22   As cognitive psychology was: U. Neisser, *Cognitive Psychology* (New York: Meredith, 1967); "History of Cognitive Psychology," compiled by the Cognitive Processes Classes, fall 1997, Psychology Department, Muskingum College, New Concord, OH, accessed September 5, 2012, http://www.muskingum.edu/~psych/ psycweb/history/cognitiv.htm; Sawyer, *Explaining Creativity*, 52–56.

22   Where before, psychologists tried: Sawyer, *Explaining Creativity*, 63–65. As Sawyer recounts, research shifted to the university level and consisted largely of controlled lab experiments, mostly with psychology graduate students, that attempted to measure memory, attention, and other basic cognitive brain functions. Much of the research design involved solving puzzles.

22   It was in the labs of Lake Forest College: Ibid., 78–79.

22   The experiments of psychologist Mihaly Csikszentmihalyi: Ibid., 87–88; Mihaly Csikszentmihalyi, *Creativity: Flow and the Psychology of Discovery and Invention* (New York: Harper Perennial, 1996).

22   "There's a certain moment": Richards and Fox, *Life*, 97.

23   At Chicago and in a series of books: Mihaly Csikszentmihalyi, *Flow: The Psychology of Optimal Experience* (New York: Harper and Row, 1990); http://en.wikipedia.org/wiki/ Special:BookSources/0060920432; Csikszentmihalyi, *The Evolving Self* (New York: Harper Perennial, 1994); Csikszentmihalyi, *Creativity*; Csikszentmihalyi, *Finding Flow: The Psychology of Engagement with Everyday Life* (Basic Books, 1998).

23   The absence of a sense of time: Sawyer, *Explaining Creativity*, 78–79; Csikszentmihalyi, *Creativity*, 111–13.

23   Back in 1959, Rollo May: Sawyer, *Explaining Creativity*, 78–79.

23   By the 1990s, advances in brain imaging: Ibid., 185.

23   It continues to be a well-funded: National Science Foundation Funding Page: Cognitive Neuroscience, accessed September

5, 2012. http://www.nsf.gov/funding/pgm_summ.jsp?pims_
id=5316.

23   Brain scans have shown: Sawyer, *Explaining Creativity*, 158–60,
195–205.

23   Because creative behaviors activate: Ibid.

24   In Renaissance Italy: Mihaly Csikszentmihalyi, "The Domain
of Creativity," in D. H. Feldman, M. Csikszentmihalyi, and H.
Gardner, eds., *Changing the World: A Framework for the Study of
Creativity* (Westport, CT: Praeger Publishers, 1994), 135–58;
Csikszentmihalyi, *Creativity*, 32–36; Sawyer, *Explaining Creativity*, 214.

24   Florence at that time: Csikszentmihalyi, *Creativity*, 32–33; Sawyer, *Explaining Creativity*, 214.

25   Interestingly, the man who observed: Csikszentmihalyi, *Creativity*, 32–36.

25   In his research and his books: Csikszentmihalyi, *Flow*; Csikszentmihalyi, *The Evolving Self*; Csikszentmihalyi, *Creativity*;
Csikszentmihalyi, *Finding Flow*.

25   Sawyer grew up in Newport News: R. Keith Sawyer, interviews
with the author, 2011–2012.

26   With his degree, Sawyer: Ibid.

26   Sawyer is now a professor: Ibid.

27   In the end, he told me: I often think about Casey's reference to
clowns. There is documentation on the OSS providing improvisation training to agents and to actors becoming agents, but I have
found nothing on clowns. Did I not hear him correctly? After all,
it was late and we'd been drinking and talking for hours. I can
see him now, sitting there, staring across the room, not looking at
me, as he was talking about people being sent out, as if he were remembering specific individuals who went out and succeeded and
others who didn't. The clown reference remains a mystery to me.

29   And like the artists who came before: Anthropologists have long
talked about the twin universes of the moral and the market, but
Michael Sandel's book *What Money Can't Buy* is the best current
analysis of the subject. Sandel takes the traditional approach of
seeing the moral—what money can't buy—in opposition to the
market: what money can buy. I'm arguing that the moral and the
market are integrated. That is, creativity bridges the two uni-

verses. It plays the role of transforming what money can't buy into what money can buy.

29   Why not, as Indian design firm: I am indebted to Sonia Manchanda, cofounder of Idiom, one of India's top design and innovation firms, for opening my eyes to this insight. Through her work on motivation, new business models, and framing the bottom of the pyramid, Manchanda is the direct heir of C. K. Prahalad.

29   People gather in a room: Alex F. Osborn, *How to "Think Up"* (McGraw-Hill, 1942); Alex F. Osborn, *Applied Imagination: Principles and Procedures of Creative Problem Solving*, Third Revised Edition (New York: Charles Scribner's Sons, 1963).

29   But as Sawyer points out: R. Keith Sawyer, *Group Genius: The Creative Power of Collaboration* (New York: Basic Books, 2007).

30   Where the older model sought: "Innovation for the Risk Averse," *Harvard Business Review*, May 2012.

31   Washington has to curry: Nicole Bullock and Vivianne Rodrigues, "Scramble to Sell US Debt as Yields Hit Record Lows," *Financial Times*, March 6, 2012, accessed September 12, 2012, http://www.ft.com/cms/s/0/04ab95a6-66e3-11e1-9e53-00144feabdc0.html#axzz20oSyTu99.

32   In 2010, for the first time: "World Demographic Profile 2012," http://www.indexmundi.com/world/demographics_profile.html, accessed September 16, 2012.

32   In forty more years: http://www.npg.org/facts/world_pop_year.htm, accessed October 18, 2012.

32   The US military's term: Carol Mase, "VUCA—A Leadership Dilemma," Free Management Library Leadership Blog, January 17, 2011, accessed September 12, 2012, http://managementhelp.org/blogs/leadership/2011/01/17/vuca-a-leadership-dilemma.

CHAPTER 3

43   commencement speech at Stanford: Steve Jobs, text and video of Stanford commencement address, June 12, 2005, posted on the Stanford Report, June 14, 2005, accessed September 4, 2012, http://news.stanford.edu/news/2005/june15/jobs-061505.html.

45   Roommates Adam Lowry: I first met Josh Handy at a Design Management Institute conference on October 25 to 27, 2010, in

Providence, Rhode Island, that I cochaired with David Butler, Design Director for Coca-Cola. During a subsequent discussion with Bill Moggridge, he presented the origin story of Method to the audience, and later, he and I talked about culture, innovation, and start-ups. Handy helped me understand that we embody knowledge as members of social groups and that it can be as important as, if not more important than, the knowledge we attain by immersing ourselves in learning new things. We focus on the "10,000 hours" of practice necessary to become an expert at something and often forget what we already know. April Dembosky, "Method in Their Madness," *Financial Times*, September 27, 2011, accessed September 4, 2012, http://www .ft.com/intl/cms/s/0/66fc9bbc-e626-11e0-960c-00144feabdc0 .html#axzz22nEQOvia.

46   "You're paying your penance": Sarah van Schagen, "An Interview with the Founders of Method Green Home-Care Products," grist.org, March 15, 2008, accessed September 4, 2012 http:// grist.org/article/fighting-dirty/

46   (their apartment actually was pretty dirty): Carlye Adler, "Method Home Cleans Up with Style and (Toxic-Free) Substance," *Time*, May 3, 2011, accessed September 4, 2012, http://business.time .com/2011/05/03/method-soap-business-profile/.

46   Lowry jumped at the opportunity: Dembosky, "Method in Their Madness."

46   The two used $90,000 that: Ibid.

46   "To us," said Lowry: Van Schagen, "An Interview with the Founders of Method Green Home-Care Products."

46   Ryan, Lowry, and Josh Handy: Josh Handy, personal interview with author, October 25–27, 2010.

47   In 2007, the company: Ibid.

47   The Method founders built: Ibid.

47   So it's not surprising: "Our Story," Method website, accessed September 4, 2012, http://methodhome.com/methodology/our-story/.

48   As a teenager, he: Walter Isaacson, *Steve Jobs* (New York: Simon & Schuster, 2011), 11.

48   When he needed parts: Ibid., 17.

50   The fact that over half: Keith N. Hampton, Lauren Sessions Goulet, Lee Rainie, Kristen Purcell, Pew Research Centers's In-

ternet & American Life Project, June 16, 2011, accessed October 19, 2012, http://pewinternet.org/Reports/2011/Technology-and-social-networks.aspx.

50   I recently talked with: Personal record, July 27, 2012.

51   It was a skill McQueen: I co-taught a class at Parsons in the fall of 2011 with Ben Lee called "Steve Jobs and Alexander McQueen: Design as Social Movement," which was attended by many students from the new MFA in Design and Society. A number had studied at Central St. Martins in London. Several talked about the great seams in McQueen's clothes and how he had apprenticed and worked at an early age in men's fashion before becoming famous in women's fashion; Andrew Bolton, *Savage Beauty* (New York: Metropolitan Museum of Art: May, 2011, fifth printing).

52   As one of China's more innovative: I first learned of Lenovo when its head of design, Yao Yingjia, came to *BusinessWeek* around the time the company was buying IBM's PC business. He wore an incredibly tailored modern "Mao" jacket that looked more Yoji Yamamoto than "Made in China" and talked more about innovation than cost and speed of delivery. I was later impressed when Lenovo asked Ziba Design, one of the smartest innovation consultancies around, to help it strategize. I then assigned a story on Lenovo's design process; "Inside Lenovo's Design Quest," *BusinessWeek*, September 24, 2006, accessed September 4, 2012, http://www.businessweek.com/stories/2006-09-24/inside-lenovos-design-quest.

53   Ziba knows how to get deep: Ziba's research for Lenovo won an IDEA award in 2006 and I wrote about it in the June 29, 2006, annual design award package. The research was showcased again in *IN: Inside Innovation*, September 2006, page 7, in "Ziba's Design's Search for the Soul of the Chinese Consumer." I first talked to Sohrab about this work in the spring of 2006.

57   Daniel Pink, author: conversation between Roger Martin and Dan Pink, March 18, 2011, on Educating the Creative Leaders of Tomorrow, put on as part of the Steelcase 360 Discussion series.

58   In 1990, MIT roboticists: Like a lot of science-fiction-loving boys in America, I grew up intrigued by robots, and I wrote about industrial robots when I got to *BusinessWeek*. But the real story

has been the disappointment with robots—how they have failed to live up to our *Star Wars* imagination. The little Roomba bots finally hit it. They opened up a new, commercial space with a robot. I believe we are now, finally, at the start of a new Robot Age. Time to see *Blade Runner* again. The history of iRobot, as well as success stories about Roomba and other products, can be found on the company's website; "Our History," iRobot, accessed September 4, 2012, http://www.irobot.com/en/us/Company/About/Our_History.aspx.

59   there's even a smallish cult: hackingroomba.com, accessed October 5, 2012.

60   Early advocates of this approach: I interviewed Lafley in 2006 for the *BusinessWeek* cover story "The Power of Design." I've talked with Claudia Kotchka, who was vice president for Design Innovation and Strategy under Lafley, many times.

60   In 2000, newly appointed CEO: Roger L. Martin, *The Opposable Mind* (Boston: Harvard Business School Publishing, 2009), 11–13, 60–61; Larry Huston and Nabil Sakkab, "P&G's New Innovation Model," March 20, 2006, *Harvard Business School Working Knowledge*, accessed October 15, 2012, http://hbswk.hbs.edu/archive/5258.html.

60   They have worked with the "NineSigma": http://www.ninesigma.com/ninesigma-overview/clients--testimonials, accessed October 3, 2012; http://innocentive.com/about-innoentive/facts-stats, accessed October 3, 2012.

60   One result was the Crest Spinbrush: "Toys and Spinning Brushes: How John Osher Found His Way to Profits," Knowledge@Wharton, November 13, 2003, accessed October 3, 2012, http://knowledge.wharton.upenn.edu/article.cfm?articleid=870; Robert Berner, "Why P&G's Smile Is So Bright," July 31, 2002, *BusinessWeek*, accessed October 3, 2012, http://www.businessweek.com/smallbiz/content/aug2002/sb2002081_2099.htm.

61   According to the *Wall Street Journal*: Michael Totty, "Paper Thin Screens with a Twist," *Wall Street Journal*, September 26, 2010, accessed September 4, 2012, http://online.wsj.com/article/SB10001424052748703470904575500342513725972.html.

62   When James Dyson went casting: Patrick Mahoney, *MachineDesign.com*, August 7, 2008, accessed October 3, 2012, http://ma

chinedesign.com/article/industrial-design-design-the-dyson-way-0807.

63   When I visited him in Toronto: Bill Buxton, interviews with the author, March 7, 2011, April 5, 2011, September 15, 2011, May 7, 2012; http://www.billbuxton.com/, accessed October 3, 2012.

65   "I put the black and white: "Van Gogh's Letters," WebExhibits, accessed September 4, 2012, http://www.webexhibits.org/vangogh/letter/20/607.htm; Debora Silverman, *Van Gogh and Gauguin: The Search for Sacred Art* (New York: Farrar, Straus and Giroux, 2000), 396.

65   Bob Dylan looked to Woody: Caspar Llewellyn Smith, *Guardian*, June 15, 2011, accessed September 4, 2012, http://www.guardian.co.uk/music/2011/jun/16/bob-dylan-woody-guthrie.

66   When asked in an interview: Cynthia McFadden, interview, ABC News, January 13, 2012, accessed September 4, 2012, http://abcnews.go.com/blogs/entertainment/2012/01/madonna-breaks-silence-on-gaga-born-this-way-controversy-2020-exclusive-tonight/.

66   has a replica of the Saturn V: Greg Klerkx, *New Scientist*, December 12, 2011, accessed at http://www.marssociety.org/home/press/news/illputmillionsofpeopleonmarssayselonmusk on October 15, 2012.

66   the powerful rocket: http://www.nasa.gov/audience/foreducators/rocketry/home/what-was-the-saturn-v-58.html, accessed October 15, 2012; http://www.time.com/time/specials/packages/article/0,28804,1910599_1910769_1910767,00.html, accessed October 18, 2012.

66   BMW bought and revived: http://www.miniusa.com/#/learn/FACTS_FEATURES_SPECS/history/storyOfMini-m, accessed October 15, 2012; http://www.topspeed.com/cars/mini/1959-2006-the-history-of-mini-ar10921.html, accessed October 15, 2012.

66   It is once again: http://www.motoringfile.com/2012/02/04/businessweek-mini-wins-big-over-smart/.

67   Paul Polak has spent: Paul Polak and Jacqueline Novogratz, founder of the Acumen Fund, are my two heroes in redesigning models to improve life in what C. K. Prahalad called the Bottom of the Pyramid. Both focus on using market forces, investments,

and start-ups to generate income and provide water, health, education, and jobs in India, Africa, and Asia. Polak talked to me about his Orissa water project over several interviews in the United States and in India in 2011 and 2012. Donald G. McNeil, Jr., "An Entrepreneur Creating Chances at a Better Life," *New York Times*, September 26, 2011, accessed September 4, 2012, http://www.nytimes.com/2011/09/27/health/27conversation .html?pagewanted=all.

67   Polak has talked to: Paul Polak Profile, "Unreasonable Network," accessed September 4, 2012, http://unreasonableinstitute.org/ profile/ppolak/; Jessie Scanlon, "Paul Polak: Global Poverty Fighter," *BusinessWeek*, January 12, 2009, accessed September 4, 2012, http://www.businessweek.com/stories/2009-01-12/paul-polak-global-poverty-fighterbusinessweek-business-news-stock-market-and-financial-advice.

67   He's seen what happens when: Paul Polak, interviews with the author, 2011 and 2012.

68   When Polak was in Bangladesh: Paul Polak, interviews with the author, 2011 and 2012; Paul Polak website, accessed September 4, 2012, http://www.paulpolak.com/html/paul.html.

68   There are now eighty-four factories: http://www.ideorg.org/Our Story/Pedaling_Out_of_Poverty.pdf, accessed October 18, 2012.

68   There are 325 million: Paul Polak, interviews with the author, 2011 and 2012; Acumen Fund, accessed September 4, 2012, http://www.acumenfund.org/investment/spring-health.html.

68   He came to understand two critical realities: Paul Polak, interviews with the author, 2011 and 2012.

69   Polak's ability to see: Ibid.

69   with the help of Indian design consultancy: Jacob Mathew, co-founder of Idiom, has worked closely with Polak on the Orissa water project from its start. Mathew is responsible for the design of the brand, the jugs that carry the water, the uniforms of the people transporting the water on bicycles, and, with Polak, the business model itself; Paul Polak, interviews with the author, 2011 and 2012.

69   "Spring Health will generate": Paul Polak, interviews with the author, 2011 and 2012; Paul Polak's blog, accessed September 4, 2012, http://blog.paulpolak.com/?m=201106.

70   The Acumen Fund, founded by: "Spring Health," The Acumen Fund, accessed September 4, 2012, http://www.acumenfund .org/investment/spring-health.html.

70   Polak hopes it can reach: Paul Polak, interviews with the author, 2011 and 2012.

71   In 2012, Jeremy Feinberg: Lisa W. Foderaro, "A New Species in New York Was Croaking in Plain Sight," *New York Times*, March 13, 2012, accessed September 4, 2012, http://www.ny times.com/2012/03/14/nyregion/new-leopard-frog-species-is-discovered-in-nyc.html; Rutgers University press release, March 2012, "Hiding in Plain Sight: Rutgers Scientist Discovers New Frog Species in New York and New Jersey," http://news .rutgers.edu/medrel/news-releases/2012/03/rutgers-ecologist-di-20120309/, accessed October 18, 2012.

71   According to CNN: Shelby Lin Erdman, "Ribbit! Frog Species Found in New York City Has a Croak of Its Own," CNN U.S., March 17, 2012, accessed October 18, 2012, http://articles.cnn .com/2012-03-17/us/us_new-york-frog-species_1_frog-species-new-frog-new-species?_s=PM:US.

71   Leslie Rissler: http://www.leslierissler.org/Leslie_Rissler/Wel come.html, accessed October 22, 2012.

71   "I've given it lots of thought": Foderaro, "A New Species in New York Was Croaking in Plain Sight."

73   Take Singapore, which: Singapore is making enormous efforts to transform itself into a creative innovative society. I was at the president's house in November 2009 as he personally gave out design awards to young Singaporeans who won special Red Dot design awards for the ICSID Design Conference. Singapore is also scoring higher and higher on different national innovation rankings. While I was there, I spoke with young designers and creatives who (especially if they had studied in Europe and the United States) still complained about the stultifying authoritarian hand of government and the obsessive consumerist culture that values Italian, French, British, and American brands, not local ones.

73   The government is making: "Singapore Government Arts and Creative Industries," accessed September 9, 2012, http://www .gov.sg/government/web/content/govsg/classic/Info_N_Policy/

ip_arts/; "Singapore Budget 2011," accessed September 9, 2012, http://www.mof.gov.sg/budget_2011/expenditure_overview/mica.html.

73   When it comes to international sports: The Online GK, October 30, 2010, http://theonlinegk.wordpress.com/2010/10/30/indian-gold-medal-winners-of-commonwealth-games-delhi/; "Indians Shine in Commonwealth Wrestling Championships," *Jagran Post*, August 8, 2011, accessed September 9, 2012, http://post.jagran.com/indians-shine-in-commonwealth-wrestling-championships-1312826371.

73   In the 2010 Commonwealth Games: Ibid.

74   There are no programs like Title IX: Usha Sharma and Sonia Manchanda, personal interviews conducted by the author, February 17, 2011.

74   Though wrestling has had a long tradition in India: Usha Sharma and Sonia Manchanda, personal interviews conducted by the author, February 17, 2011; Roshan Kumar, "Women Wrestlers Cry Disparity," *Telegraph*, Calcutta, India, March 14, 2012, accessed September 9, 2012, http://www.telegraphindia.com/1120314/jsp/bihar/story_15246222.jsp#.UE1eia60J8F.

74   Usha Sharma, a police officer: Usha Sharma, personal interview conducted by the author on February 17, 2011.

74   Sharma's wrestling academy: Ibid.

74   Sons are preferred over daughters: prepared by Krishan S. Nehra, Senior Foreign Law Specialist: "Sex Selection and Abortion: India," Law Library of Congress website, accessed September 9, 2012, http://www.loc.gov/law/help/sex-selection/india.php.

74   Haryana has 830 girls: "Seven Brothers: India's Skewed Sex Ratio," April 7, 2011, *Economist*, accessed October 18, 2012, http://www.economist.com/node/18530371.

74   "I have always wanted": Sonia Manchanda, interview with the author, February 17, 2011.

75   Manchanda and Idiom launched: Sonia Manchanda, interview with the author, February 17, 2011; Sarah Jacob, "Idiom: The Dream Merchants Who Turn Your Dreams into Business Ideas," *India Times*, December 3, 2011, accessed September 9, 2012, http://articles.economictimes.indiatimes.com/2011-12-03/news/30471796_1_dreams-idiom-volunteers.

75 Manchanda brought in: I was invited to participate in the second phase of the Dream:IN conference by Carlos Teixeira, a professor at Parsons The New School of Design. It was held at Idiom headquarters in Bangalore from February 16 to 19, 2011, and I joined with students and businesspeople to craft business plans for three of the aspirations that came out of the interviews that took place months earlier. They all revolved around education. A second Dream:IN conference was held in Brazil, Carlos's home country, in the summer of 2012, www.fastcodesign.com/1663223/looking-to-realize-dreams-in-india-with-new-business-models.

75 "More education" was probably the most: I was present in India, participating at the Dream:IN conference, when the major themes were presented after the video narratives were collected and collated.

76 They already have $50,000: Sonia Manchanda, interview with author, autumn 2011.

76 a country where more than 65 percent: Kaushik Basu, "India's Demographic Dividend," BBC News, July 25, 2007, accessed September 10, 2012, http://news.bbc.co.uk/2/hi/south_asia/6911544.stm.

77 Keith Richards and Mick Jagger: Keith Richards and James Fox, *Life* (New York: Little, Brown and Company, 2010).

77 Today, the business degree: National Center for Education Statistics, Fast Facts, http://nces.ed.gov/fastfacts/display.asp?id=37, accessed September 16, 2012; Michelle Megna, Schools .com, 5 most popular undergraduate majors of 2011 entry-level hires, http://www.schools.com/articles/most-popular-degrees-employers.html, accessed September 16, 2012.

78 In a talk given to a freshman: Bill Deresiewicz, article adapted from address to students at Stanford University, May 2010, http://chronicle.com/article/What-Are-You-Going-to-Do-With/124651/.

78 I recently had lunch with someone: This informal talk occurred in the spring of 2012.

80 3M was a pioneer of this strategy: Paul D. Kretkowski, "The 15 Percent Solution," Wired, January 23, 1998, accessed October 19, 2012, http://www.wired.com/techbiz/media/news/1998/01/9858.

80 Google has had a 20 percent: James Whittaker, March 13, 2012,

"Why I Left Google," *JW on Tech* blog, accessed October 18, 2012, http://blogs.msdn.com/b/jw_on_tech/archive/2012/03/13/ why-i-left-google.aspx.

82   Steve Jobs was a walker: Bill Moggridge, personal conversation, winter 2012. I knew Bill Moggridge for many years when he was at IDEO. He interviewed me for his book *Designing Media*, October 2010, for my transition from print to digital journalism at *BusinessWeek*. A pioneer in designing the first laptop, the Grid, which is now in the permanent collection of MoMA, and even more important, the interaction between people and computers, Moggridge was a giant in the design world. I was fortunate to get to know him a bit personally when he moved to New York to head the Cooper-Hewitt National Design Museum.

83   The late Pulitzer Prize–winning: Donald M. Murray, "Real Writers Don't Burn Out: Making a Writing Apprenticeship Last a Lifetime," keynote address, National Writers' Workshop, Hartford, CT, April 1, 1995 (posted by Bill Mitchell, Poynter, August 25, 2002), accessed October 19, 2012, http://www.poynter .org/uncategorized/2063/real-writers-dont-burn-out-making-a-writing-apprenticeship-last-a-lifetime/.

CHAPTER 4

85   Charles Adler loved music: I first met Charles in Chicago at the Design Research Conference put on by the IIT Institute of Design from October 24 to 26, 2011. He gave a presentation at that conference about the origins of Kickstarter. I interviewed him November 9, 2011, and he gave a talk in my Parsons course Design at the Edge the following spring (2012).

86   Chen was living in New York: Charles Adler, personal interviews with the author. Charles Adler presentation in the author's Parsons course Design at the Edge, spring 2012; Carlye Adler, "How Kickstarter Became a Lab for Daring Prototypes and Ingenious Products, *Wired*, March 18, 2011, accessed September 11, 2012, http://www.wired.com/magazine/2011/03/ff_kickstarter/2/.

86   All transactions are handled: Yancey Strickler, "Amazon Payments and US-Only" Kickstarter Blog post, October 3, 2009,

http://www.kickstarter.com/blog/amazon-payments-and-us-only, accessed September 11, 2012.

86   On one level Chen, Strickler: Charles Adler: personal interviews with the author. Charles Adler presentation in the author's Parsons course Design at the Edge, spring 2012.

87   The first Kickstarter projects: Ibid.

87   Between its launch in 2009 and October 2012: http://www.kick starter.com/help/stats, accessed October 4, 2012.

87   which had an operating budget: http://www.arts.gov/about/bud get/appropriationshistory.html, accessed October 19, 2012.

87   A campaign for new watches: "Transform Your iPod Nano into the World's Coolest Multi-Touch Watches with TikTok + LunaTik by Scott Wilson and MINIMAL," Kickstarter campaign site, http://www.kickstarter.com/projects/1104350651/tiktok-lunatik-multi-touch-watch-kits.

87   San Francisco–based studio raised: "Doublefine Adventure," Kickstarter campaign page, accessed September 11, 2012, http://www.kickstarter.com/projects/doublefine/double-fine-adventure?ref=live.

88   JOBS Act, new legislation: Mark Landler, "Obama Signs Bill to Promote Start-Up Investments," *New York Times*, April 5, 2012, accessed September 11, 2012, hhttp://www.nytimes .com/2012/04/06/us/politics/obama-signs-bill-to-ease-investing-in-start-ups.html; Ryan Caldbeck, "How the JOBS Act Could Change Startup Investing Forever," *TechCrunch*, March 16, 2012, accessed September 11, 2012, http://techcrunch .com/2012/03/16/crowdfundingstartups/.

88   We all hold a number: I am deeply indebted to my wife, Leslie M. Beebe, who, until recently, was Professor of Sociolinguistics at Teachers College, Columbia University, for showing me the significance of the concept of Framing. She pointed me to the deep well of research on the subject and introduced me to the work of her friend Deborah Tannen, Professor of Linguistics at Georgetown University. Leslie highlighted the status function of Framing as well—how it can present you as "one up" or "one down." How you frame yourself can reinforce or help change your status.

88   Gregory Bateson, a British anthropologist: Gregory Bateson, *Steps to an Ecology of Mind: Collected Essays in Anthropology, Psychi-*

atry, Evolution, and Epistemology (Chicago: University of Chicago Press, 1972).

89   He painted side by side: "Pablo Picasso's Cubism Period—1909 to 1912," http://www.pablopicasso.org/cubism.jsp, accessed September 9, 2012.

89   But then he went further: "MoMA Featured Works: Picasso Guitars, 1912–1914," *Still Life with Guitar*: http://www.moma .org/interactives/exhibitions/2011/picassoguitars/featured-works/35.php; *Guitar*: http://www.moma.org/interactives/ex hibitions/2011/picassoguitars/featured-works/36.php, accessed September 9, 2012. Holland Cotter, "When Picasso Changed His Tune," *New York Times*, February 10, 2011, accessed October 20, 2012, http://www.nytimes.com/2011/02/11/arts/ design/11picasso.html?pagewanted=all&_r=0.

89   In 2011, the Museum of Modern Art: Andrea Kirsh, "Picasso, Music and Negative Space: The Guitars at MoMA," posted on The Art Blog, February 21, 2011, accessed September 9, 2012, http://www.theartblog.org/2011/02/picasso-music-and-negative-space-the-guitars-at-moma/. I was fortunate enough to see the Picasso Guitars exhibit at MoMA, and it was a powerful visual trigger in understanding the role of reframing in creativity.

89   Ninety-four years later, a young artist: I first met Marla Allison in Santa Fe when she won the first Innovation Award given out by SWAIA, the Southwestern Association for Indian Arts, at the annual Indian Market for her portrait of her mother in 2009. She came to New York to give presentations in my Design at the Edge classes in 2011 and 2012. Although I waited too long and missed the chance to buy the painting of her mother, I do have another hanging on my wall that portrays the hills of her Laguna Pueblo where potters go to collect their clay.

90   Erving Goffman, a Canadian-born: Erving Goffman, *Interaction Ritual: Essays on Face-to-Face Behavior* (New York: Anchor Books, 1967); Erving Goffman, *Frame Analysis: An Essay on the Organization of Experience* (London: Harper and Row, 1974).

90   But today, thanks to outrage at the 1 percent: Kevin Roose, "A Blow to Pinstripe Aspirations," *New York Times*, November 11, 2011, accessed September 11, 2012, http://dealbook.nytimes

.com/2011/11/21/wall-st-layoffs-take-heavy-toll-on-younger-workers/.

91 Chemotherapy is a difficult therapy: I heard Irish Maliq speak at the GE HealthCare Conference Health by Design in 2009. Over the next couple of years she told me more about Memorial Sloan-Kettering's innovative approach to the delivery of chemotherapy, and in 2011, I got to attend a presentation given by the students in which they analyzed the MSK Infusion Center project. I had a chance to talk to them about their reframing process afterward.

92 In the end, the neighborhood chemo: Reframing the delivery of medicine from something given to patients in a massive, centralized, and faraway hospital to something offered in a small, friendly, local place is one of the most important trends taking place in the medical industry. The MSKCC Brooklyn Infusion Center, with its street-front gallery and storefront access, is in the vanguard of this trend.

92 One of my first cover stories for *BusinessWeek*: Bruce Nussbaum, "I Can't Work This Thing!" *BusinessWeek*, April 28, 1991, accessed September 11, 2012, http://www.businessweek.com/stories/1991-04-28/i-cant-work-this-thing. (The title of this article has been changed on *BusinessWeek's* website. The title of the print version was "I Can't Make the !@#&%! Thing Work.")

94 In his 2004 book *Don't Think of an Elephant!*: George Lakoff, *Don't Think of an Elephant!* (White River Junction, VT: Chelsea Green Publishing Co, 2004).

95 IBM, for example, found itself: Lisa DiCarlo, "How Lou Gerstner Got IBM to Dance," *Forbes*, November 11, 2002, accessed September 11, 2012, http://www.forbes.com/2002/11/11/cx_ld_1112gerstner.html.

95 IBM was able to reframe: Ira Sager, "How IBM Became a Growth Company Again," *BusinessWeek*, December 9, 1996, accessed September 13, 2012, http://www.businessweek.com/1996/50/b35051.htm.

95 Kodak was slow to rethink: There were many stories written about the decline of Kodak, but one of the best was by Steve Hamm, who wrote this insightful piece for *IN:Inside Innovation* in February 2007, while I was editor. Steve was perhaps too op-

timistic about the possibility of resurrection; Steve Hamm, "Kodak's Moment of Truth," *BusinessWeek: IN,* February 18, 2007, accessed September 15, 2012, http://www.businessweek.com/stories/2007-02-18/kodaks-moment-of-truth.

96 Global corporations like Unilever: Kay Johnson/Xa Nhon, "Marketing: Selling to the Poor," *Time,* April 17, 2005, accessed September 11, 2012, http://www.time.com/time/magazine/article/0,9171,1050276,00.html.

98 As for "gifts," they like Groupon: I first realized that "gifting" had tremendous social and economic importance when I was in the Philippines as a Peace Corps volunteer. Filipino society was then, in the late 1960s, based on *utang na loob,* or "debt from inside." It's a reciprocity culture and people feel the obligation to pay their economic and social debts. Japan, for all its modernity, is built on a similar system. Understanding the deep meaning of the gift and what "free" often implies is a key skill in comprehending social media and modern life in the United States, Asia, Europe, Latin America, and the Middle East.

99 I was on a jury: I saw the KNOCK Project presentation in Toronto on April 19, 2011, at the Microsoft Design Expo held at the Design Exchange, an event put on by Bill Buxton for his students. Susan Gorbet and Spencer Saunders were the creators of KNOCK.

101 Lenovo, the Chinese computer company: Chuck Salter, "Protect and Attack: Lenovo's New Strategy," *Fast Company,* November 22, 2011, accessed September 11, 2012, http://www.fastcompany.com/1793529/protect-and-attack-lenovos-new-strategy; *Bloomberg News,* "China Rural Incomes Rising Most Since '84 Show Lure for Job-Seeking Obama," January 20, 2011, accessed October 20, 2012, http://www.bloomberg.com/news/2011-01-20/china-rural-incomes-rising-most-since-84-show-lure-for-job-seeking-obama.html.

101 By making the box representative: Ibid.

101 At a Design Management Conference in 2010: I organized and cochaired with David Butler, the head of design for Coca-Cola, the M5+D=nV conference in Providence, Rhode Island, October 25 to 27, 2010. It was based on five macro changes—the rise and fall of nations, the rise and fall of generations, social media,

urbanization, and sustainability—that design can have a huge impact on and generate new value.

101   "You know you are going to": Kathleen Taylor spoke at the M5+D=nV conference organized by the 2010 Design Management Conference.

102   Several years ago, I asked my students where: Creativity and innovation probably happen across all eras and time frames, but they appears to cluster and spike at certain specific moments in certain specific places, often cities. Florence in the 1400s, as Mihaly Csikszentmihalyi observed, for example. London in the sixties. Silicon Valley in the nineties. New York City, where I currently live, is in the midst of a huge surge in creativity and entrepreneurship. But why? How does this come about? I asked my students at Parsons to begin mapping their inspirations as they live, work, study, and create in New York. Where do they go to be challenged? Who do they actually connect with? What neighborhoods are the "hottest" and why? It's a project that I hope will grow so we can eventually have creativity maps of Shanghai and Berlin, Paris and São Paolo, Chicago and Stuttgart. It may even help us promote creativity in our personal lives.

104   At a Design Indaba Conference: The Design Indaba Conference (February 26 to 28, 2010) was my first time in Cape Town. It was an inspiring, stimulating series of days that showed just how creative South Africa truly is. Ravi Naidoo is the organizer of this annual conclave, and his energy, intellect, and global network of connections are powers to behold. Martha missed an opportunity, because she was stuck in the same old frame.

104   The *Economist* is turning: http://www.economistgroupmedia .com/products/eiu/audience/, accessed October 20, 2012.

105   Pat Pruitt, a metalsmith from New Mexico: Each year in August, SWAIA, or the Southwest Association for Indian Arts, judges hundreds of works by Native artists at the Indian Market. Winners are announced, and they, along with hundreds of other artists, sell their art under a cloud of white tents set up in the streets of Santa Fe. If you want a special work of art, you have to line up in the dead of night. I got to Pat Pruitt's tent at 4 a.m. to buy his concho belt in the summer of 2011. I use it in my classes to

illustrate the power of aura and the need for strong engagement between people and objects.

105   There are billboards for Calvin Klein: Giselle Tsirulnik, "Calvin Klein Activates Billboards with QR Codes Pushing Mobile Video Ad," *Mobile Marketer*, July 29, 2012, accessed September 12, 2011, http://www.mobilemarketer.com/cms/news/advertis ing/6933.html.

105   The art world, too, is embracing: I've worked with Paola Antonelli in the World Economic Forum's Global Agenda Council on Design and she has spoken in my Design at the Edge class several times. Her exhibits at the MoMA, from her first, *Mutant Materials*, showcased the engagement between people and design. Paola is a champion of the designer's role as the bridge between science and society, technology and humanity.

105   But Antonelli's exhibits: Personal record of presentation, Paola Antonelli.

105   In their 1999 book *The Experience Economy*: J. Pine and J. Gilmore, *The Experience Economy* (Boston: Harvard Business School Press, 1999).

106   GBN, the scenario-planning arm: Aging Asia: Economic and Social Implications of Rapid Demographic Change in China, Japan, and Korea. February 26–27, 2009, conference, co-organized by Shorenstein APARC and the Global Aging Program at the Stanford Center on Longevity, accessed October 14, 2012, http:// asiahealthpolicy.stanford.edu/events/5501.

107   According to GBN's Bulletin: http://asia healthpolicy.stanford .edu.events/5501; "Asia Is Getting Older—How About Wiser?" *GBN Bulletin*, Summer 2009, 5.

107   Nearly all the experts on population: http://longevity1.stanford.edu/ AgingAsia/Pages/Day2Links.html, accessed October 20, 2012.

107   Lisa K. Solomon is a principal: I first met Lisa Solomon at the Design Management Institute conference on October 25–27, 2010, in Providence, RI, that I cochaired with David Butler, the Vice President of Global Design, The Coca-Cola Company. Her presentation, Scenario Planning: Preparing for the Future of Design, blew the audience away.

109   Arthur C. Clarke was the quintessential: Dylan Tweney, "Sci-Fi Author Predicts Future by Inventing It," *Wired*, May 25, 2011,

http://www.wired.com/thisdayintech/2011/05/0525arthur-c-clarke-proposes-geostationary-satellites/; Gerald Jonas, "Arthur C. Clarke, Author Who Saw Science Fiction Become Real, Dies at 90," *New York Times*, March 19, 2008, accessed September 11, 2012, http://www.nytimes.com/2008/03/19/books/19clarke.html?pagewanted=all; Neil McLain, CSBE, "Arthur C. Clarke: The Father of Satellite Communications," Society of Broadcast Engineers Chapter 24, Inc., accessed September 11, 2012, http://www.ctiinfo.com/SatControl/ComTrack/InclinedOrbitTutorial/appndxb.htm.

110   The geostationary satellite concept: http://www.britannica.com/EBchecked/topic/524891/satellite-communication/224536/Development-of-satellite-communication, accessed October 20, 2012.

110   It took that long: http://www.boeing.com/defense-space/space/bls/deltaHistory.html, October 20, 2012.

110   with the second of his "Three Laws": Arthur C. Clarke, "Hazards of Prophecy: The Failure of Imagination" in *Profiles of the Future*, (London: Scientific Book Club, Book Club Edition, 1962).

110   In a piece entitled: Jonas, "Arthur C. Clarke . . . Dies at 90"; Tweney, "Sci-Fi Author Predicts Future by Inventing It."

111   One company that was able to successfully: Bombardier Company history, accessed September 15, 2012, http://www.bombardier.com/en/corporate/about-us/history?docID=0901260d8001dffa.

112   If Kodak executives: see earlier note in this chapter regarding Kodak, Hamm, "Kodak's Moment of Truth."

113   Had Philips, the Dutch corporate consumer giant: http://www.flatpanelshd.com/news.php?subaction=showfull&id=1303112824, accessed October 15, 2012.

113   Umpqua Bank, one of Oregon's oldest: I spoke with Sohrab Vossoughi, founder of Portland-based Ziba Design, about his work with Umpqua. As we walked through the streets of Portland to see the Umpqua branches, I was surprised and delighted by the first thing you see at the branch—the water dish for dogs. Sohrab got the culture right and engaged Umpqua's customers strongly and emotionally even before they entered the bank.

114   In a piece for the *New York Times Opinionator*: Tina Rosenberg, "For Healthy Aging, a Late Act in the Footlights," August 15,

2012, accessed September 15, 2012, http://opinionator.blogs
.nytimes.com/2012/08/15/for-healthy-aging-a-late-act-in-the-
footlights/.

114   "We live in a society that's very acute-care based": Ibid.

CHAPTER 5

117   West, the CEO of the Boston-based: Continuum is one of the
great design and innovation consultancies, and I've had many
talks with Harry West and the firm's founder Gianfranco Zac-
cai over the years. It is most famous for its research and development of the Swiffer mop for Procter & Gamble, now one of
P&G's most profitable brands. But Continuum's work in financial
services, medical products, and brand strategy is widely known
throughout the United States and Europe. The last time I saw
Gianfranco was when we were both speakers at the ICSID World
Design Conference in Singapore in 2009. Harry West has presented in my Design at the Edge class, and I talk with him often;
this story is based on my interview with West on March 20, 2012,
in New York City.

117   Dr. Ruben Rausing is usually credited: "Who We Are: Our
Legacy," Tetra Pak USA website, accessed September 10, 2012,
http://www.tetrapak.com/us/whoweare/heritage/pages/default
.aspx.

117   Continuum then launched: Harry West interview, March 20,
2012, New York City.

118   In a recent *Harvard Business Review*: "Life's Work: Richard
Serra," *Harvard Business Review*, March 2010, accessed September 13, 2012, http://hbr.org/2010/03/lifes-work-richard-serra/
ar/1.

119   Though there are countless ways of playing: My conversations
with Katie Salen, a graduate of the Rhode Island School of
Design in game design, and my reading of her book, have had
a huge impact on how I understand the creative process. Salen
introduced me to the idea of "magic circles" and connected the
engaged interaction of gaming to the educational philosophy of
John Dewey, who talked about "learning by doing." Salen is helping to remake the face of public education; she has set up three

public schools—two in Chicago, one in New York—that team up teachers with game designers to build an exciting learning experience for students; Institute of Play, http://www.instituteofplay .org/about, accessed September 17, 2012.

119 Though scholars are in disagreement: Marilyn Yalom, *Birth of the Chess Queen* (New York: HarperCollins, 2004), 3; David Shenk, *The Immortal Game* (New York: Doubleday, 2006), 16–20.

119 "This was a war game": Shenk, *The Immortal Game.*

119 According to military strategist: Max Boot, *War Made New* (New York: Gotham Books, 2006), 122.

120 In 2002, General Tommy Franks: GlobalSecurity.org, http:// www.globalsecurity.org/military/ops/internal-look.htm, accessed September 13, 2012.

120 Internal Look was also used: Mark Mazzetti and Thom Shanker, "U.S. War Game Sees Perils of Israeli Strike Against Iran," *New York Times*, March 19, 2012, accessed September 13, 2012, http:// www.nytimes.com/2012/03/20/world/middleeast/united-states-war-game-sees-dire-results-of-an-israeli-attack-on-iran.html?_r=2.

120 The outcome of this particular game: Ibid.

120 In 1905, German economist: Max Weber, *The Protestant Ethic and the Spirit of Capitalism* (New York: Taylor & Francis e-Library, 2005; orig. pub. Allen and Unwin, 1930), 116.

120 American economist Frank Knight: Ben Lee, former Provost and Professor of Philosophy and Anthropology at the New School, first brought my attention to the significance of Chicago economist Frank Knight's work in linking business and economics to uncertainty, games, and play. Knight's work provided a fundamental underpinning for much of this book; Frank Hyneman Knight, *The Ethics of Competition* (New Brunswick, NJ: Transaction Publishers, 2009, originally published in 1935 by Harper and Brothers and Allen & Unwin), 38.

121 When Barbara Walters asked: "Famous Montessori Students, Parents, and Supporters," http://www.michaelolaf.net/ google.html, accessed September 13, 2012; ABC News, "A Fascinating Group," http://abcnews.go.com/Entertainment/ story?id=309165&page=1#.UEu57q60J8E.

121 "We both went to Montessori School": ABC News, "A Fascinating Group."

121   Montessori alums include: Peter Sims, "The Montessori Mafia," *Wall Street Journal*, April 5, 2011, accessed September 13, 2012, http://blogs.wsj.com/ideas-market/2011/04/05/the-montessori-mafia/.

121   Other entrepreneurs with educational backgrounds: "Google Logo, Founders Spell Success: Montessori," HispanicBusiness .com, August 31, 2012, accessed September 13, 2012, http://www.hispanicbusiness.com/2012/8/31/google_logo_founders_spell_success_montessori.htm.

121   Paul Graham, the founder of Y Combinator: Randall Stross, *The Launch Pad: Inside Y Combinator, Silicon Valley's Most Exclusive School for Startups* (New York: Portfolio, 2012); http://paulgraham.com/bio.html.

121   Biz Stone, cofounder of Twitter: CMO.com, "Twitter Creator Biz Stone Chats with Adobe CMO Lewnes at Digital Summit 2012, http://m.cmo.com/leadership/twitter-creator-biz-stone-chats-adobe-cmo-lewnes-digital-summit-2012, accessed September 13, 2012.

121   "Being playful, less structured": Melissa Korn and Amir Efrati, "Master of 'Biz' Returns to School," *Wall Street Journal*, September 1, 2011, accessed September 13, 2012, http://online.wsj.com/article/SB10001424053111904009304576533010574207444.html?mod=WSJ_hpp_MIDDLE_Video_Third.

122   In fact, bringing Katie Andresen: Bruce Nussbaum, "INside Innovation—Lessons Learned from Open-Sourcing Innovation," *BusinessWeek*, May 25, 2006, accessed September 13, 2012, http://www.businessweek.com/stories/2006-05-25/inside-innovation-lessons-learned-from-open-sourcing-innovation-dot.

123   In 2008, I went to a conference: I attended Serious Play in 2008 in Pasadena. Chee Pearlman organized this extraordinary conference that highlighted for me the importance of play to creativity. A second conference, Radical Craft, foreshadowed the resurgence of the Maker culture. Chee was a key mentor when I began covering design in the early nineties. Chee is founder of Chee Company, which provides design and innovation content to conferences, websites, magazines, books, and all of today's "platforms." Chee also runs the Curry Stone Design Prize awards program.

123   Not long ago, Craig Wynett called me: personal record of conversation September 12, 2011.

124   In times of relative stability: US Department of Defense News Transcript, February 12, 2002, http://www.defense.gov/transcripts/transcript.aspx?transcriptid=2636, accessed September 13, 2012.

124   Chad Hurley, the cofounder of YouTube: I was in the audience for this interview, and asked questions afterward. Bill Moggridge held an extraordinary series of public conversations with prominent designers while he was running the Cooper-Hewitt National Design Museum. The audiences were small, perhaps fifty to seventy-five people, and the talk was casual and insightful, with plenty of time for questions. I attended the conversation with Chad Hurley.

125   Walter Isaacson's biography: Walter Isaacson, *Steve Jobs* (New York: Simon & Schuster, 2011), 346.

126   "The stage, the screen": J. Huizinga, *Homo Ludens: A Study of the Play-Element in Culture* (London: Routledge, 1949), 10; conversations with Katie Salen.

127   And finally, don't be afraid: Harry West, interview with Bruce Nussbaum, March 20, 2012, New York City.

129   Adam Bosworth, the founder: http://keas.com/blog/keas -introduces-the-power-of-play-in-a-new-online-health-and-well ness-social-game-that-delivers-unprecedented-employee-health- engagement-rates/, accessed October 20, 2012.

129   "If you want to get people": Adam Bosworth, "Web of Games," *TechCrunch*, June 18, 2011, accessed September 13, 2012, http:// techcrunch.com/2011/06/18/web-of-games/.

129   When he headed: Adam Bosworth, interview with Eric Schonfeld, "Google Health Creator Adam Bosworth on Why It Failed: It's Not Social," *TechCrunch*, June 24, 2011, accessed October 22, 2012, http://techcrunch.com/2011/06/24/google-health- bosworth-social/.

129   Eric Bailey, Aimee Jungman, and: Eric Bailey, Aimee Jungman, and Thomas Sutton, "Google Health's Failure to Bring Meaning to Data," *Design Mind*, June 28, 2011, accessed September 13, 2012, http://designmind.frogdesign.com/blog/google-health-s- failure-to-bring-meaning-to-data.html.

130   Google Health attracted: Steve Lohr, "Google to End Health Records Service After It Fails to Attract Users," *New York Times*, June 24, 2011, accessed September 13, 2012, http://www.ny

times.com/2011/06/25/technology/25health.html; Marshall Kirkpatrick, "Google Health: Why It's Ending and What It Means," ReadWriteWeb, June 24, 2011, accessed September 13, 2012, http://www.readwriteweb.com/archives/google_health_why_its_ending_what_it_means.php.

130   Bosworth didn't let the failure: Elizabeth Gudrais, "Playing with Health," *Harvard Magazine*, May to June 2012, accessed September 13, 2012, http://harvardmagazine.com/2012/05/playing-with-health.

130   With more than one third: Centers for Disease Control and Prevention, http://www.cdc.gov/obesity/data/adult.html, accessed October 20, 2012.

130   More than 100,000 people: Gudrais, "Playing with Health."

130   Of the eight thousand employees: Miguel Helft, "Getting Healthy Is All Fun and Games for Keas," *CNN Money*, March 9, 2012, accessed September 13, 2012, http://tech.fortune.cnn.com/2012/03/09/keas-health-social-gaming/.

130   In twelve-week pilot programs: http://keas.com/success/, accessed September 13, 2012.

131   Virgin HealthMiles: Helft, "Getting Healthy Is All Fun and Games for Keas"; http://us.virginhealthmiles.com/Pages/Home.aspx, accessed September 13, 2012; RedBrick Health, https://www.redbrickhealth.com/, accessed September 13, 2012.

131   SuperBetter is one of the most creative additions: SuperBetter website, https://www.superbetter.com/, accessed September 13, 2012.

131   Though a passionate believer: Bruce Feiler, "She's Playing Games with Your Life," *New York Times*, April 17, 2012, accessed September 13, 2012, http://www.nytimes.com/2012/04/29/fashion/jane-mcgonigal-designer-of-superbetter-moves-games-deeper-into-daily-life.html.

132   And so SuperBetter's brightly designed: http://www.superbetter.com/, accessed September 13, 2012.

132   "Play is a voluntary activity": Huizinga, *Homo Ludens*, 7.

132   And a great deal in shirts: Gilt.com, http://www.gilt.com, accessed September 13, 2012.

133   It raised another $138 million: Evelyn M. Rusli, "Gilt Groupe Valued at Roughly $1 Billion," *New York Times*, May 9, 2011, accessed September 13, 2012, http://dealbook.nytimes.com/2011/05/09/

gilt-groupe-valued-at-roughly-1-billion/; Tomio Geron, "Gilt Groupe Raises $138 Million Led by Japan's Softbank," *Forbes*, May 9, 2011, September 13, 2012, http://www.forbes.com/sites/tomiogeron/2011/05/09/gilt-groupe-raises-138-million-led-by-japans-softbank/.

133   At its height, 80 million: Oliver Chiang, "FarmVille Players Down 25% Since Peak, Now Below 60 Million," *Forbes*, October 15, 2012, accessed October 18, 2012, http://www.forbes.com/sites/oliverchiang/2010/10/15/farmville-players-down-25-since-peak-now-below-60-million/.

133   And millions more people: Adam Holisky, "World of Warcraft Subscriber Numbers Dip 100,000 to 10.2 Million," joystiq.com, February 9, 2012, accessed September 13, 2012, http://wow.joystiq.com/2012/02/09/world-of-warcraft-subscriber-numbers/.

134   Bob Greenberg, CEO: Bob Greenberg told me this story when I was at *BusinessWeek*. He is one of the most inspiring creators I've ever met. Dressed in New York black and wearing cool silver bracelets, balding with long silver hair, he cuts a definitive figure across the creative landscape of both coasts. He started R/GA doing creative work for the movies and reinvents the company every nine years. In 2012, Greenberg looked at the fast growth of R/GA to over one thousand and decided to split it up into teams of 150 each and add new services, from product design to strategic consulting; Nike+, http://judgeseyesonly.com/nikeplus.html, accessed September 13, 2012.

134   So R/GA designed a website: personal interview with Greenberg; http://judgeseyesonly.com/nikeplus_video.html.

135   There are many different kinds of games: personal interview with Katie Salen, June 6, 2011, New York City; Katie Salen and Eric Zimmerman, *Rules of Play: Game Design Fundamentals* (Boston: MIT Press, 2004), 80–83.

136   Friedrich Froebel: One of the legends of product design, Tucker Viemeister first presented this connection between progressive education and design and creativity at a DMI conference that I cochaired with David Butler, the design director of Coca-Cola. It was eye-opening, and I asked him to present it to my class, which he did, in the spring of 2011. He's the only designer I know who was named after a car, the Tucker, which his father

designed. http://www.friedrichfroebel.com/, accessed October 20, 2012.

136 the progressive education movement expanded: http://en.wikipedia.org/wiki/Montessori_education, accessed September 13, 2012; http://www.montessori-ami.org, accessed September 13, 2012; http://en.wikipedia.org/wiki/History_of_Waldorf_schools, accessed September 13, 2012.

136 In 2007, Katie Salen: interview with Katie Salen, June 6, 2011.

136 received a MacArthur Foundation grant: http://www.instituteofplay.org/about, accessed September 13, 2012.

137 Salen puts on a weeklong summer: interview with Katie Salen, June 6, 2011, Institute of Play, http://www.instituteofplay.org/work/projects/mobile-quest, accessed September 13, 2012.

137 Perhaps that is why 72 percent: Video Game Voters, http://videogamevoters.org/pages/top_10_gamer_facts/, accessed September 13, 2012.

137 StarCraft II: John Gaudiosi, "Major League Gaming Wraps Record-Breaking 2011 Season with Over $600,000 in Cash and Prizes," GamerLive.TV, November 21, 2011, accessed September 13, 2012, http://www.gamerlive.tv/article/major-league-gaming-wraps-record-breaking-2011-season-over-600000-cash-and-prizes; Gunnar Technology Eyewear, http://www.gunnars.com/events/gunnar-mlg-providence-national-championships/, accessed September 13, 2012.

138 Re-Mission is a game: Re-Mission website, http://www.re-mission.net/, accessed September 13, 2012.

138 The game was created by HopeLab: "About HopeLab," http://www.hopelab.org/about-us/, accessed September 13, 2012.

138 According to a study conducted: Pamela M. Kato, Steve W. Cole, Andrew S. Bradlyn, and Brad H. Pollock, "A Video Game Improves Behavioral Outcomes in Adolescents and Young Adults with Cancer: A Randomized Trial," *Pediatrics*, vol. 122, no. 2, August 1, 2008, accessed September 13, 2012, http://pediatrics.aappublications.org/content/122/2/e305.full.

138 "A game designer": Edutopia, "Big Thinkers," www.edutopia.org/digital-generation-katie-salen-video?page=1, accessed September 13, 2012; http://en.wikipedia.org/wiki/SimCity, accessed October 21, 2012.

139 Humans vs. Mosquitoes: http://humansvsmosquitoes.com/back ground/, accessed September 13, 2012; Lauren Graham, "Climate Conversations—Can a Game Combat Malaria?" Alert-Net, July 17, 2012, accessed October 20, 2012, http://www .trust.org/alertnet/blogs/climate-conversations/can-a-game-combat-malaria/.

139 It was designed by students: Ibid.

140 In 1485, Leonardo da Vinci: http://www.flyingmachines.org/ davi.html, accessed September 13, 2012.

140 It was not only a beautiful work: http://en.wikipedia.org/wiki/ Ornithopter, accessed September 13, 2012.

140 in 1959, a wealthy British businessman: Aza Raskin, "Wanna Solve Impossible Problems? Find Ways to Fail Quicker," *Fast Company*, accessed October 15, 2012, http://www.fastcodesign .com/1663488/wanna-solve-impossible-problems-find-ways-to-fail-quicker.

141 Aeronautics engineer Paul B. MacCready, however: Fiddlers Green, http://www.fiddlersgreen.net/models/aircraft/Gossamer-Albatross.html, accessed September 13, 2012.

141 saw that the game itself was flawed: Raskin, "Wanna Solve Impossible Problems? Find Ways to Fail Quicker."

141 MacCready was able to fly three: Ibid.

142 This is what the four founding members: http://www.ucbcom edy.com/podcasts/ucbtny, accessed September 13, 2012.

143 In a 2011 interview: Ibid.

143 This ran counter to the way: Ibid.

143 as well as an acclaimed training: "About Us," Upright Citizens Brigade Theatre, accessed September 15, 2012, http://newyork .ucbtheatre.com/about.

145 As Greenberg learned: In creating Nike+, Greenberg built one of the first product/experience ecosystems outside Apple. He was a decade ahead of others in understanding that a key business strategy should be to build not just products or services but an entire ecosystem within which the consumers participate and create their own identities.

CHAPTER 6

147   It was 3:44 in the morning: "SpaceX Launch—NASA," http://
       www.nasa.gov/exploration/commercial/cargo/spacex_index
       .html, accessed September 7, 2012.

147   The Dragon capsule was free: Clara Moskowitz, "SpaceX
       Launches Private Capsule on Historic Trip to Space Station," May
       22, 2012; http://www.space.com/15805-spacex-private-capsule-
       launches-space-station.html, accessed September 7, 2012.

147   Just days after the launch: "Space X," accessed September 7, 2012,
       http://www.nasa.gov/exploration/commercial/cargo/spacex_in
       dex.html.

148   This flight was, after all: "Elon Musk, CEO and Chief Designer,"
       http://www.spacex.com/elon-musk.php, accessed September 7,
       2012.

148   Many know Elon Musk: Ibid.

148   pretty much at the nadir: Encyclopedia of World Biography,
       http://www.notablebiographies.com/news/Li-Ou/Musk-Elon
       .html#b, accessed September 7, 2012.

148   By 2002, eBay realized: http://news.cnet.com/2100-1017-941964
       .html, accessed September 7, 2012.

148   In 2002, Musk became the CEO: Margaret Kane, "eBay picks
       up PayPal for $1.5 Billion," CNET News, July 8, 2002; http://
       www.notablebiographies.com/news/Li-Ou/Musk-Elon.html#b,
       accessed September 7, 2012.

149   A year later he founded a second: Will Oremus, "Tesla's New
       Electric Car Is Practical and Affordable, as Long as You're Rich,"
       Slate, June 20, 2012, accessed September 7, 2012, http://www
       .slate.com/blogs/future_tense/2012/06/20/tesla_model_s_new_
       electric_car_is_practical_affordable_for_the_rich.html.

149   But Musk said in his celebratory: Ibid.

149   In 2007, just before the biggest: Gabriel Sherman, "The End of
       Wall Street as They Knew It," *New York* magazine, February 5,
       2012, accessed September 7, 2012, http://nymag.com/news/fea
       tures/wall-street-2012-2/index3.html.

149   Historically, banks never accounted: Gillian Tett, *Financial
       Times* US editor and author of *Fool's Gold*, shared this informa-
       tion at a March 10, 2010, presentation at Columbia University;

Sherman, "The End of Wall Street as They Knew It."

150   the majority of business school graduates: personal interview with Roger Martin; Rakesh Khurana, *From Higher Aims to Hired Hands: The Social Transformation of American Business Schools and the Unfulfilled Promise of Management as a Profession* (Princeton, NJ: Princeton University Press, 2010), 328–31, 349.

150   But by the end of the century: Ibid.

150   Top bankers received astonishing: Linda Anderson, "MBA Careers: Financial Services—A Breadth of Opportunity," *Financial Times*, January 29, 2007, accessed September 7, 2012, http://www.ft.com/intl/cms/s/2/3baa68a4-ad5a-11db 8709-0000779e2340,dwp_uuid=991cbd66-9258-11da-977b 0000779e2340.html#axzz22nEQOvia.

150   When *BusinessWeek* ran: January 31, 2000, issue, cover story by Michael Mandel.

151   An inequality gap: Sam Pizzigati, "Happy Days Here Again, 21st Century–Style," Institute for Policy Studies, March 13, 2012, accessed September 7, 2012, http://www.ips-dc.org/blog/happy_ days_here_again_21st_century-style.

151   Alice Waters's groundbreaking organic: "About Chez Panisse," http://en.wikipedia.org/wiki/Chez_Panisse, accessed September 7, 2012.

152   Just as important, Gen Y: I joined Parsons in 2008, and I am indebted to my Parsons students for these and other insights into Gen Y culture.

152   You can pay about a hundred bucks: TechShop website, http:// www.techshop.ws/, accessed September 7, 2012.

152   *Make* magazine, launched in 2005: http://makezine.com/maga zine/, accessed September 7, 2012.

153   The Faires celebrate "arts, crafts": http://makerfaire.com/new york/2012/index.html, accessed September 7, 2012.

153   Generation Y, on the other hand: interviews with Kelsey Meuse in my classroom and after graduation.

154   Bombarded with as many as five thousand: Louise Story, "Anywhere the Eye Can See, It's Likely to See an Ad," *New York Times*, January 15, 2007, accessed September 5, 2012, http:// www.nytimes.com/2007/01/15/business/media/15everywhere .html?pagewanted=all.

154 sales of vinyl albums: http://www.businesswire.com/news/home/
20120105005547/en/Nielsen-Company-Billboard%E2%80%99s-
2011-Music-Industry-Report, accessed October 12, 2012; "It's
Official—Vinyl Sales Up 39 percent in 2011," Digital Music News, January 4, 2011, http://www.digitalmusicnews.com/
permalink/2012/120104vinyl/, accessed September 8, 2012; "Back
to Black," *Economist*, August 20, 2011, http://www.economist
.com/node/21526296, accessed September 8, 2012.

154 The generations who grew up in a fast-food: interviews with students, 2009 through 2012.

155 To anyone under thirty: Parsons classroom discussions, 2008
through 2012.

155 You can go to Etsy, Fab.com, or the Vintage Typewriter Shoppe:
http://imitationobjects.com/art/restored-vintage-typewriters-
fab-com/, accessed September 7, 2012; http://www.etsy.com/
shop/TheAntikeyChop, accessed September 7, 2012; http://
www.vintagetypewritershoppe.com/Vintage_Typewriters_25
.html, accessed September 7, 2012.

156 people are using film: David Graham, "Developing into a Thing of
the Past," thestart.com, April 3, 2008, accessed September 8, 2012,
http://www.thestar.com/living/article/409180/; http://fwd.chan
nel5.com/gadget-show/gadget-news/polaroid-600-one-camera-
gets-reissued, accessed September 8, 2012; http://www.chemie
.unibas.ch/~holder/slr690.htm, accessed September 8, 2012.

156 folks eager to develop: http://www.photokaboom.com/photogra
phy/learn/tips/014_how_to_develop_film.htm, accessed September 8, 2012; http://www.kodak.com/global/en/consumer/educa
tion/lessonPlans/darkroom/fullCourse.shtml, accessed September
8, 2012; Jessica Schira, "How to Develop Film in a Darkroom," accessed September 8, 2012, http://www.ehow.com/how_4466203_
develop-film-darkroom.html.

156 Jennie Dundas and Alexis Miesen: http://www.bluemarble
icecream.com/, accessed September 8, 2012.

157 Their Brooklyn-based company: Deborah L. Cohen, "Entrepreneurs Get the Scoop on Success," *Reuters*, May 19, 2010, accessed
September 8, 2012, http://www.reuters.com/article/2010/05/19/
us-column-cohen-bluemarble-idUSTRE64I4BM20100519.

157 made with milk from grass-fed: http://travel.yahoo.com/ideas/

best-ice-cream-spots-in-the-u-s-.html?page=all, accessed October 15, 2012; http://www.slowfoodnyc.org/program/snail_ap proval/awardee/blue_marble_ice_cream; accessed October 15, 2012; http://www.bluemarbleicecream.com/, accessed September 8, 2012.

157 Launched in 2007, just as: Cohen, "Entrepreneurs Get the Scoop on Success"; http://www.bluemarbleicecream.com/, accessed September 8, 2012; Jen Carlson, "Jenny and Alexis: Blue Marble Ice Cream," *Gothamist*, October 7, 2008, accessed September 8, 2012, http://gothamist.com/2008/10/07/jennie_ and_alexis_blue_marble_ice_c.php; http://www.triposo.com/ poi/N__832243111, accessed September 8, 2012.

157 Miesen says the name: http://www.bluemarbleicecream.com/, accessed September 8, 2012; http://cosmiclog.nbcnews.com/_ news/2012/1/2/07/15755286-40-years-later-apollo-17s-blue- marble-leaves-a-mark-on-our-memory?lite, accessed December 10, 2012.

157 "want to know the origins": Carlson, "Jenny and Alexis: Blue Marble Ice Cream."

157 "to connect with the faces, cultures": Ibid.

158 The stores use only biodegradable: Ibid.

158 Coming full circle to Miesen's original: Zoe Schlanger, "Blue Marble to Open Rwanda's First Ice Cream Shop," *Gothamist*, May 28, 2012, accessed September 8, 2012, http://gothamist .com/2010/05/28/blue_marble_to_open_rwandas_first_i.php; Leslie Goldman, "Ice Cream Entrepreneurs Bring Sweetness— and Jobs—to Rwanda," *Oprah*, August 2011, accessed September 8, 2012, http://www.oprah.com/spirit/Helping-Rwanda-Blue- Marble-Ice-Cream-Jennie-Dundas.

158 Babette, a successful: http://www.babettesf.com/, accessed September 8, 2012.

159 And the most difficult restaurant: http://www.news.com.au/ breaking-news/denmarks-noma-retains-best-restaurant-title/ story-e6frfku0-1226343377226, accessed September 8, 2012; http://noma.dk/reservations/, accessed September 8, 2012.

160 General Electric is about as global: David Wessel, "Big U.S. Firms Shift Hiring Abroad," *Wall Street Journal*, April 19, 2011, accessed September 8, 2021, http://online.wsj.com/article/SB1

00014240527487048217045762707836611823972.html; http://
www.gereports.com/immelt-on-60-minutes-how-growing-
global-markets-translate-into-american-jobs/.

160   In the rush away from manufacturing: Geoffrey Colvin and Katie
        Brenner, "GE Under Siege," *CNN Money*, October 15, 2008, ac-
        cessed September 8, 2012, http://money.cnn.com/2008/10/09/
        news/companies/colvin_ge.fortune/index2.htm.

160   Only after the financial crash of 2007: Stephanie Clouser, "GE's
        Jeff Immelt Offers 10 Ways to Be More Competitive," *Business
        First*, March 21, 2012, accessed September 8, 2012, http://www
        .bizjournals.com/louisville/blog/2012/03/immelt-shares-his-
        thoughts-on-what-it.html.

160   GE is spending $1 billion: http://www.gereports.com/ge-
        accepting-applications-for-480-new-appliance-manufacturing-
        jobs-in-louisville, accessed September 8, 2012; http://www
        .gereports.com/ge-invests-1-billion-in-high-tech-plants-to-
        open-u-s-manufacturing-jobs/, accessed September 8, 2012.

160   Water heaters and washing machines: Ed Crooks, "GE
        Takes a $1 Billion Risk in Bringing Jobs Home," *Financial
        Times*, April 2, 2012, accessed October 20, 2012, http://www
        .ft.com/intl/cms/s/0/21a46546-78f1-11e1-88c5-00144feab49a
        .html#axzz29rvIaGu3; Pamela Coyle, "GE Heats Up Louis-
        ville, KY's Appliance Park with New Investment," Business Cli-
        mate.com, accessed September 8, 2012, http://businessclimate
        .com/kentucky-economic-development/ge-heats-louisville-kys-
        appliance-park-new-investment.

160   Lower prices for technology: Lisa Harrington, "Is U.S. Manufac-
        turing Coming Back?" *Inbound Logistics*, August 2011, accessed
        September 8, 2012, http://www.inboundlogistics.com/cms/ar
        ticle/is-us-manufacturing-coming-back/.

161   So Chip Blankenship, chief executive: Jason Heiner, "Be-
        hind GE's Feature-Heavy Refrigerator, a Lean Manufacturing
        Strategy," Smart Planet, March 22, 2012, accessed Septem-
        ber 8, 2012, http://www.smartplanet.com/blog/smart-takes/
        behind-ges-feature-heavy-refrigerator-a-lean-manufacturing-
        strategy/24564.

161   Soon many that were stamped: http://www.gereports.com/an-
        inside-look-at-ges-manufacturing-prowess/, accessed September

8, 2012; Harrington, "Is U.S. Manufacturing Coming Back?"

161   Christine Furstoss, Technical Director: personal interviews with
      Christine Furstoss and Beth Comstock, spring 2012; http://
      ge.geglobalresearch.com/about/technology-directors/christine-
      furstoss/, accessed September 8, 2012.

161   The company found that with: personal interviews with Chris-
      tine Furstoss and Beth Comstock; http://www.3dprinter.net/3d-
      printing-make-ge-jet-engines-lighter-efficient, accessed Septem-
      ber 8, 2012.

161   Typically these devices are made: personal interviews with GE's
      Beth Comstock and Christine Furstoss.

161   Printing them is much easier: personal interviews with GE's Beth
      Comstock and Christine Furstoss; David H. Freedman, "Layer
      by Layer," *Technology Review*, January/February 2012, accessed
      September 8, 2012, http://www.technologyreview.com/featured
      story/426391/layer-by-layer/.

161   The process begins with spreading: personal interviews with GE's
      Beth Comstock and Christine Furstoss.

162   "We are seeing the convergence": http://www.ge.com/audio_
      video/ge/news_events/a_look_inside_ge_garages.html, accessed
      September 8, 2012.

162   Recently GE "printed" out: personal interviews with GE's Beth
      Comstock and Christine Furstoss.

162   Two of Chrysler's most popular: "Imported from Detroit" and
      "The Things We Make, Make Us," http://www.youtube.com/
      watch?v=Mi0SbrrGaiw, accessed September 8, 2012.

162   When Daimler AG: http://www.leftlanenews.com/videos-
      dieter-zetsche-stars-in-new-chrysler-ads.html#, accessed Sep-
      tember 8, 2012.

162   Its ads focused on German technology: http://www.youtube
      .com/watch?v=SKL254Y_jtc, accessed September 8, 2012.

162   Amy Turn Sharp and her husband: Dean Narciso, "Recalls Spur
      Old-Fashioned Toy Carving," *Columbus Dispatch*, February 7,
      2009, accessed October 21, 2012, http://www.dispatch.com/
      content/stories/local/2009/02/07/woodshop.ART_ART_02-
      07-09_B3_LVCR16U.html; "Little Alouette: Wood for Wee
      Ones," http://www.littlealouette.com/story, accessed September
      8, 2012.

163   Joe, who grew up in Britain: "Little Alouette: Wood for Wee Ones."

163   the high prices of vintage: http://www.ebay.com/sch/i .html?_nkw=Old+German&_sacat=717&_odkw=german&_ osacat=717, accessed September 8, 2012.

163   Amy and Joe began: "Little Alouette: Wood for Wee Ones."

163   Kids "require so little": "Little Alouette Has Big Plans for Baby Toys," post by Walker, *Columbus Underground*, October 19, 2009, September 8, 2012, http://www.columbusunderground.com/ little-alouette-has-big-plans-for-baby-toys.

163   And for every purchase: http://www.etsy.com/listing/48850576/ little-alouette-wee-wood-moustache, accessed September 8, 2012.

163   In 2007, the Sharps opened: http://www.etsy.com/shop/littleal ouette, accessed September 8, 2012.

163   Rattles and teethers: http://www.littlealouette.com/blocks/ safari-free-play-block-set, accessed September 8, 2012.

164   Sharp is an Etsy mom: http://en.wikipedia.org/wiki/Mompre neur, accessed September 8, 2012; http://www.wisegeek.com/ what-is-a-mompreneur.htm, accessed September 8, 2012.

164   They set up shop, literally: "Little Alouette Has Big Plans for Baby Toys."

164   millions who shop there: http://mywifequitherjob.com/etsy stores/, accessed September 8, 2012.

164   Rob Kalin dropped out: Teri Evans, "Creating Etsy's Handmade Marketplace," *Wall Street Journal*, March 30, 2010, accessed September 8, 2012, http://online.wsj.com/article/SB10001424052 702304370304575152133860888958.html?mod=WSJ_hp_edi torsPicks; http://www.nyu.edu/about/university-initiatives/en trepreneurship/nyu-ventures/success-stories/etsy.html, accessed September 8, 2012.

164   His real passion: Evans, "Creating Etsy's Handmade Market-place."

164   In 2005, he got two: http://www.nyu.edu/about/university initiatives/entrepreneurship/nyu-ventures/success-stories/etsy .html, accessed September 8, 2012.

164   Headquarters were set up: http://www.etsy.com/help/arti cle/486/, accessed September 8, 2012.

164   Etsy has 875,000 sellers: I am indebted to Larry Keeley of Doblin, part of the Monitor Group of consultants, for first showing me the importance of platforms and platform innovation. Before we had Etsy, Amazon, or eBay, Larry was telling his clients that they should be seeking big, disruptive innovations via their business platforms. Today, with Apple's huge success, platform innovation is the new orthodoxy; Jenna Wortham, "Etsy Raises $40 Million for International Expansion," *New York Times* Bits Blog, May 29, 2012, accessed September 8, 2012, http://bits.blogs.nytimes.com/2012/05/09/etsy-raises-40-million-for-international-expansion/.

164   And there is talk: Inc. staff, "Rob Kalin Out as Etsy CEO," Inc., July 21, 2011, accessed September 8, 2012, http://www.inc.com/articles/201107/rob-kalin-steps-down-as-etsy-ceo.html.

164   Etsy has been able: Wortham, "Etsy Raises $40 Million for International Expansion."

165   The year before, the company: Ibid.

165   the site charges sellers: http://www.etsy.com/sell?ref=so_sell, accessed September 8, 2012; Alex Williams, "That Hobby Looks Like a Lot of Work," *New York Times*, December 16, 2009, accessed September 8, 2012, http://www.nytimes.com/2009/12/17/fashion/17etsy.html.

165   Etsy's user base: Williams, "That Hobby Looks Like a Lot of Work."

165   Martha Stewart Media gave: http://thecraftsdept.marthastewart.com/2010/11/vote-for-the-winners-in-the-holiday-craft-sale.html, accessed September 8, 2012.

165   The toy company then moved: http://www.littlealouette.com/FAQ, accessed September 8, 2012.

165   Swiss Miss, Cool Mom Picks: http://www.littlealouette.com, accessed September 8, 2012.

166   The site does not track: Williams, "That Hobby Looks Like a Lot of Work."

166   But over the last year: Wortham, "Etsy Raises $40 Million for International Expansion."

166   "multinational of one": Tim Brown used this term when he spoke at my class in the spring of 2012.

167   Dan Provost, an interaction: http://www.crunchbase.com/person/dan-provost, accessed September 8, 2012.

167   He is one of the millions: http://www.kickstarter.com/projects/

danprovost/glif-iphone-4-tripod-mount-and-stand, accessed September 8, 2012.

167   "Because of its small form": Ibid.

167   he joined with his friend: http://tomgerhardt.com/, accessed September 8, 2012.

167   neither had any direct: Jenna Wortham, "A Web Edge for Makers of Real Stuff," *New York Times*, April 20, 2011, accessed September 8, 2012, http://www.nytimes.com/2011/04/21/technology/21make .html?_r=1&scp=2&sq=Glif&st=cse&pagewanted=all.

167   it needed a tripod: http://www.kickstarter.com/projects/danpro vost/glif-iphone-4-tripod-mount-and-stand, accessed September 8, 2012.

167   The two began by sketching: http://www.therussiansusedapen cil.com/post/2794775825/idea-to-market-in-5-months-making-the-glif, accessed September 8, 2012.

167   "From the beginning, it was clear": Ibid.

168   They quickly came up: Ibid.

168   The price of design software: Ibid.

168   A Brooklyn-based start-up: http://www.makerbot.com, accessed September 8, 2012.

169   If you break your paper: Ibid.

169   Services like Shapeways: http://www.therussiansusedapencil .com/post/2794775825/idea-to-market-in-5-months-making-the-glif, accessed September 8, 2012.

169   And because 3-D printing: Ibid.

169   Apple stores offer one-on-one: http://www.apple.com/retail/ learn/one-to-one/, accessed September 8, 2012.

169   Instructables.com, an online: http://www.instructables.com, accessed September 8, 2012.

169   You can learn how: Ibid.

170   After Provost and Gerhardt: http://www.therussiansusedapen cil.com/post/2794775825/idea-to-market-in-5-months-making-the-glif, accessed September 8, 2012.

170   "People launching projects": personal interviews with Charles Adler, one at a research conference in Chicago at the IIT Institute of Design on November 9, 2011, the other in the spring of 2012 in my class Design at the Edge.

170   "too slick": "An Atom-Based Product, Developed in Bits," post

by G.F., *Economist's* Babbage blog, October 6, 2010, accessed September 8, 2012, http://www.economist.com/blogs/babbage/2010/10/small-scale_production.

170 The second featured: http://www.therussiansusedapencil.com/post/2794775825/idea-to-market-in-5-months-making-the-glif, accessed September 8, 2012.

170 they raised the entire $10,000: Ibid.

170 And the money kept: Ibid.; "An Atom-Based Product, Developed in Bits."

170 They received $137,417: http://www.kickstarter.com/projects/danprovost/glif-iphone-4-tripod-mount-and-stand?ref=live, accessed October 20, 2012.

170 The top sites for generating: http://www.therussiansusedapencil.com/post/2794775825/idea-to-market-in-5-months-making-the-glif, accessed September 8, 2012.

170 Provost and Gerhardt went searching: Ibid.

171 They are designing and making: http://www.studioneat.com/products/cosmonaut/, accessed September 9, 2012.

172 "Seeing a motorcycle about to leave": Matthew B. Crawford, *Shop Class as Soulcraft: An Inquiry into the Value of Work* (New York: Penguin, 2009), 4–5.

CHAPTER 7

177 As a member of the seminal: The story about Beats was told to me by Bob Brunner over many phone conversations and presented to my Design at the Edge class at Parsons.

177 a producer who has worked: The Richest.org, accessed September 4, 2012, http://www.therichest.org/entertainment/dr-dre-net-worth/; Andrew J. Martin, "Headphones with Swagger (and Lots of Bass)," *New York Times*, November 19, 2011, http://www.nytimes.com/2011/11/20/business/beats-headphones-expand-dr-dres-business-world.html?_r=1&pagewanted=1&partner=rss.

177 He's also known to be a perfectionist: Recording Connection Audio Institute, accessed September 4, 2012, http://www.recordingconnection.com/artists/dr-dre.

177 "People aren't hearing all the music": http://ellauniverse.blogspot

.com/2010/07/hear-all-music-with-beats-by-dr-dre.html, accessed October 20, 2012.

177 In his three decades making music: Zack O'Malley Greenburg, "Hip-Hop's Wealthiest Artists 2011," *Forbes*, accessed September 4, 2012, http://www.forbes.com/sites/zackomalleygreen burg/2011/03/09/the-forbes-five-hip-hop-wealthiest-artists/.

178 The man Dre teamed up with: personal interview with Bob Brunner; personal record from Brunner's presentation in my spring 2012 Design at the Edge class.

180 In 2011, the Designer Fund: http://designerfund.com, accessed September 5, 2012.

180 About 180,000 master's degrees: National Center for Educational Statistics, accessed September 5, 2012, http://nces.ed.gov/fastfacts/display.asp?id=37; Kelly Holland, "Is It Time to Retrain B-Schools?" *New York Times*, March 14, 2009, accessed September 5, 2012, http://www.nytimes.com/2009/03/15/business/15school.html.

180 Courses in entrepreneurialism are among: Personal interviews with deans of a number of business schools in North America.

180 Harvard Business School, a longtime: http://www.hbs.edu/entrepreneurship/, accessed September 5, 2012.

181 Richard Florida has long discussed: Richard Florida's website, accessed September 5, 2012, http://www.creativeclass.com/richard_florida/books/the_rise_of_the_creative_class; Richard Florida, *The Rise of the Creative Class: And How It's Transforming Work, Leisure, Community and Everyday Life* (New York: Basic Books, 2002).

181 A 2012 study for the Center: http://nycfuture.org/content/articles/article_view.cfm?article_id=1306, accessed September 5, 2012.

181 Since 2007, local venture capital: http://www.crunchbase.com/company/ia-ventures, accessed September 5, 2012.

181 Tumblr CEO David Karp describes: Dana Rubinstein, "On Bloomberg's Alley Versus Valley Designs, Tumblr's David Karp Explains That the Flavors Are Different," *Capital New York*, February 16, 2012, accessed September 5, 2012, http://www.capitalnewyork.com/article/politics/2012/02/5280012/bloombergs-alley-versus-valley-designs-tumblrs-david-karp-explains-.

181 There were around a dozen tech incubators: http://www.quora.com/What-are-the-top-startup-incubators-accelerators-and-

startup-coworking-spaces-in-NYC, accessed September 4, 2012.

181   Go to a meeting of NY Creative Interns: personal interview with
      Emily Miethner, who presented in my class; http://nycreativein
      terns.com/about/, accessed September 5, 2012.

181   Even New York Mayor: William Glaberson, "Life After Salomon
      Brothers," *New York Times*, October 11, 1987, accessed Septem-
      ber 5, 2012, http://www.nytimes.com/1987/10/11/business/life-
      after-salomon-brothers.html.

181   and whose financial data company: http://www.bloomberg.com/
      company/, accessed September 5, 2012.

182   In 2011, he set up a contest: Oliver Staley and Henry Gold-
      man, "Cornell, Technion Are Chosen by New York City to
      Create Engineering Campus," Bloomberg.com, December 19,
      2011, accessed September 5, 2012, http://www.bloomberg.com/
      news/2011-12-19/cornell-university-said-to-be-chosen-by-new-
      york-for-engineering-campus.html.

182   The word "pivot" is often: Lizette Chapman, " 'Pivoting' Pays Off
      for Tech Entrepreneurs," *Wall Street Journal*, April 26, 2012, ac-
      cessed September 5, 2012, http://online.wsj.com/article/SB1000
      1424052702303592404577364171598999252.html; Adam Tratt,
      "Our Startup's Pivot: Three Important Lessons We Learned,"
      GeekWire, July 19, 2012, accessed September 5, 2012, http://
      www.geekwire.com/2012/pivot-boss-3-lessons-learned/.

182   Instagram, for example, started: M. G. Siegler, "A Pivotal
      Pivot," *Tech Crunch*, November 8, 2010, accessed September 5,
      2012, http://techcrunch.com/2010/11/08/instagram-a-pivotal-
      pivot/.

182   Burbn didn't succeed: Ibid.

183   In his book *What Money Can't Buy*: Michael J. Sandel, *What
      Money Can't Buy: The Moral Limits of Markets* (New York: Farrar,
      Straus and Giroux. 2012).

184   Sales of the headphones: personal interview with Bob Brunner;
      http://beatsbydre.com/, accessed September 5, 2012.

184   And the design, too, was key: http://ellauniverse.blogspot
      .com/2010/07/hear-all-music-with-beats-by-dr-dre.html,
      accessed September 5, 2012.

185   Dr. Dre has amplified the Beats: http://www.squidoo.com/
      coolest-headphones#module155639586, accessed September 5,

2012; http://wireless-headphones-review.toptenreviews.com/, accessed September 5, 2012.

185   in August 2012, the NPD Group: Ben Arnold, "From Compton with Love: Beats by Dre Makes Some Noise at the Olympics in London," August 6, 2012, accessed October 15, 2012, https://www.npdgroupblog.com/from-compton-with-love-beats-by-dre-makes-some-noise-at-the-olympics-in-london/.

185   A Beats store just opened: http://beatsbydre.com/, accessed September 5, 2012.

185   in 2012, Beats Electronics bought: Tom Cheredar, "Hear That? Beats Electronics Officially Buys MOG," *Venture Beat*, July 2, 2012, accessed September 5, 2012, http://venturebeat.com/2012/07/02/beats-mog/; Mike Snider, "Beats Electronics Acquires MOG Music Service," *USA Today*, July 2, 2012, accessed September 5, 2012, http://content.usatoday.com/communities/technologylive/post/2012/07/beats-electronics-accquires-mog-music-service/1#.UCMmGHDgJT4.

185   that year Beats also dropped: Antony Bruno, "Jimmy Iovine on the Beats/HTC Deal: 'The Record Industry Must Make the Transition to Phones Globally,'" Billboard.biz, August 11, 2012, accessed September 5, 2012, http://www.billboard.biz/bbbiz/industry/digital-and-mobile/jimmy-iovine-on-beats-htc-deal-the-record-1005314212.story.

185   "secular epiphanies": The first time I heard the term "secular epiphany" was when my colleague and co-teacher Ben Lee used it in reference to Max Weber and Émile Durkheim and their work on the religious origins of capitalism.

185   In his 1933 essay: Jun'ichiro Tanizaki, *In Praise of Shadows*, tr. Thomas J. Harper and Edward G. Seidensticker (Stony Creek, CT: Leete's Island Books, 1977), 16.

186   Walter Benjamin argued: Walter Benjamin, "The Work of Art in the Age of Mechanical Reproduction," in *Illuminations*, Hannah Arendt, ed. (New York: Schocken, 1968).

187   When Steve Jobs returned: Walter Isaacson, *Steve Jobs* (New York: Simon & Schuster, 2011), 348–57.

187   Designers Jonathan Ive and Danny Coster: Peter Burrows, "Who Is Jonathan Ive? The Man Behind Apple's Design Magic," *IN: Inside Innovation*, September 2006. The "jelly bean story" was

told to Peter Burrows at the Radical Craft Conference at the Art Center College of Design, Pasadena, in 2006, and reported on in *IN* magazine, a quarterly magazine inside *BusinessWeek*, which I founded that year and edited. I assigned and edited the story, suggesting to John Byrne, Managing Editor of *Business-Week* at the time, that he should get Burrows to this conference.

187 Ive's team of designers: Burrows, "Who Is Jonathan Ive?"; http://designmuseum.org/exhibitions/online/jonathan-ive-on-apple/imac-1998, accessed September 5, 2012.

187 Next the team traveled: Burrows, "Who Is Jonathan Ive?"; http://designmuseum.org/exhibitions/online/jonathan-ive-on-apple/imac-1998, accessed September 5, 2012; Janet Abrams, "Radical Craft/The Second Art Center Design Conference," http://www.core77.com/reactor/04.06_artcenter.asp, accessed September 5, 2012.

187 Ive then spent yet more: Burrows, "Who Is Jonathan Ive?"

188 They also designed a beautiful: Neil Hughes, "Book Details Apple's 'Packaging Room,' Steve Jobs' Interest in Advanced Cameras," *Apple Insider*, January 24, 2012, accessed September 5, 2012, http://www.appleinsider.com/articles/12/01/24/book_details_apples_packaging_room_interests_in_advanced_cameras_.html; Yonu Heisler, "Inside Apple's Secret Packaging Room," *Network World*, January 24, 2012, accessed September 5, 2012, http://www.networkworld.com/community/blog/inside-apples-secret-packaging-room.

188 The iMac's launch in 1998: http://www.youtube.com/watch?v=0BHPtoTctDY, accessed September 5, 2012; http://designmuseum.org/design/jonathan-ive, accessed September 5, 2012; John Webb, "10 Success Principles of Apple's Innovation Master Jonathan Ive," *Innovation Excellence*, April 30, 2012, accessed September 5, 2012, http://www.innovationexcellence.com/blog/2012/04/30/10-success-principles-of-apples-innovation-master-jonathan-ive/.

188 In a 2006 interview with Peter Burrows: Burrows, "Who Is Jonathan Ive?"

188 Apple is the world's largest company: https://www.google.com/finance?client=ob&q=NASDAQ:AAPL, accessed October 17, 2012.

189 As Walter Isaacson's biography: Isaacson, *Steve Jobs*.

189   The iTunes app acts: http://www.apple.com/itunes/, accessed
       September 5, 2012.

189   face-to-face dimension: Benjamin, "The Work of Art in the Age
       of Mechanical Reproduction."

190   Boeing has begun using new: accessed September 5, 2012, http://
       www.boeing.com/commercial/787family/background.html/; ac-
       cessed September 5, 2012, http://www.boeing.com/commercial/
       aeromagazine/articles/qtr_4_06/article_04_2.html.

190   IBM has moved into: Jessi Hempel, "Crowdsourcing," Sep-
       tember 24, 2009, accessed September 5, 2012, http://www
       .businessweek.com/stories/2006-09-24/crowdsourcing; http://
       www-07.ibm.com/services/ph/portfolios/ITS/its_s_cs_c_na
       tionalcity.html, accessed September 9, 2012; http://www.cnbc
       .com/id/17169877?__source=vty, accessed September 9, 2012.

191   Corning is developing new: http://9to5mac.com/2012/06/04/
       corning-announces-slim-flexible-willow-glass-video/, accessed
       September 5, 2012; www.apple.com/about/job-creation/, ac-
       cessed September 9, 2012; http://en.wikipedia.org/wiki/Go
       rilla_Glass, accessed September 9, 2012.

191   From its founding in 1939: In the spring of 2012, I assembled a
       panel of six retired HP engineers and researchers who'd worked
       there from the early glory days through the company's decline,
       and spent two days talking with them in order to understand the
       culture of HP and how it had changed.

191   advanced degrees in electrical engineering: Lee Fleming, "Find-
       ing the Organizational Sources of Technological Breakthroughs:
       The Story of Hewlett-Packard's Thermal InkJet," *Industrial and
       Corporate Change*, vol. 11, no. 5, 1059–84 (Oxford University
       Press, 2002); "Case Study: Spitting Image," *Economist*, Septem-
       ber 19, 2002, accessed September 10, 2012, http://www.econo-
       mist.com/node/1324685.

192   "HP Labs was a wonderful place": Fleming, "Finding the Orga-
       nizational Sources of Technological Breakthroughs."

192   "I bore easily": Ibid.

192   "very far, very fast": Ibid.

192   In 1978, Vaught and Donald: Ibid.

192   From the beginning of what: http://en.wikipedia.org/wiki/
       Dot_matrix_printer, accessed September 5, 2012; http://eight

iesclub.tripod.com/id325.htm, accessed September 5, 2012.

192 Dot-matrix printers were "impact printers": Stan Retner, "History of Inkjet Printers Development," *Toner Cartridge Depot*, November 21, 2007, accessed September 5, 2012, http://blog .tonercartridgedepot.com/2007/11/21/history-of-inkjet-printers-development/.

192 Printing was slow and loud: http://en.wikipedia.org/wiki/Dots_per_inch, accessed September 5, 2012.

192 In fact, the joke going: personal interviews with the six retired HP engineers I talked with in Portland, Oregon, in the spring of 2012.

193 For most of its early history: Ibid.; Frank Cloutier, "Building One of the World's Largest Technology Businesses (and How to Have Fun and Profit from Your Hobbies)," presentation at MIT, March, 2, 2004, accessed at http://techtv-dev .mit.edu/videos/15930-building-one-of-the-world-s-largest-technology-businesses-and-how-to-have-fun-and-profit-from-your-home.

193 "We weren't the largest": Cloutier, "Building One of the World's Largest Technology Businesses."

193 And yet on Christmas Eve: Fleming, "Finding the Organizational Sources of Technological Breakthroughs; "Case Study: Spitting Image."

193 as Vaught caught sight: Fleming, "Finding the Organizational Sources of Technological Breakthroughs."

193 "Inventors just don't go home": Ibid.

193 "if you think about it": Ibid.

193 (Because of this explosive process): Thomas Kraemer, "Printing Enters the Jet Age," *American Heritage Invention and Technology*, Spring 2001, vol. 6, no. 4, 18–27; accessed September 5, 2012, http://tomsosu.blogspot.com/2012/02/history-of-hp-inkjet-printers-in.html.

194 "They had tremendous fun": Alan G. Robinson and Sam Stern, *Corporate Creativity* (San Francisco: Berrett-Koehler Publishers, Inc., 1997, 1998), 161.

194 In three months' time: "Case Study: Spitting Image."

194 The process promised to be fast: Ibid.; Robinson and Stern, *Corporate Creativity*, 165–66.

194   "carried the ball in selling": Fleming, "Finding the Organizational Sources of Technological Breakthroughs."

194   "Vaught doggedly pursued his interest": "Case Study: Spitting Image."

194   "Because its inner workings": Robinson and Stern, *Corporate Creativity*, 162.

195   "phreatic reaction": "Case Study: Spitting Image."

195   "the worst period of his life": Robinson and Stern, *Corporate Creativity*, 163.

195   Meanwhile, Frank Cloutier: Kraemer, "Printing Enters the Jet Age."

195   Cloutier's role was to support: Ibid.

195   HP at that time had a lot of wanderers: personal interviews with the six retired HP engineers I talked with in Portland, Oregon, in the spring of 2012.

195   The key was to keep the labs: Ibid.

196   Cloutier found what he was looking for: Robinson and Stern, *Corporate Creativity*, 165.

196   Thanks to help from colleagues: Ibid., 163–64.

196   "technological performance and manufacturing": Frank L. Cloutier, "Managing the Development of a New Technology," *Hewlett-Packard Journal*, May 1985, 39.

196   "One was unrivaled communications": Ibid.

196   In a 2004 speech at MIT: Cloutier, "Building One of the World's Largest Technology Businesses."

197   "As you think about visions": Ibid.

197   It took the work of hundreds: Robinson and Stern, *Corporate Creativity*, 165.

197   it wasn't until 1984: http://www.hp.com/hpinfo/abouthp/histnfacts/museum/imagingprinting/0011/index.html, accessed September 5, 2012.

197   That year, HP also released: http://www.hp.com/hpinfo/abouthp/histnfacts/museum/imagingprinting/0018/index.html, accessed September 5, 2012.

197   Only in 1988 did HP: Kraemer, "Printing Enters the Jet Age."

197   its first color inkjet: http://www.hpmuseum.net/divisions.php?did=4, accessed September 15, 2012.

197   printer went on to become: Sergio G. Non, "Will Merger Hurt

HP's Printing Biz?" ZDNet, February 8, 2002, accessed September 5, 2012, http://www.zdnet.com/news/will-merger-hurt-hps-printing-biz/120630.

197   About 300 million have been: HP release, "Twenty Years of Innovation," http://www.hp.com/hpinfo/newsroom/press_kits/2008/deskjet20/bg_deskjet20thannivtimeline.pdf.

197   In 2011, HP's Imaging: Ibid.

197   A month before: Robinson and Stern, *Corporate Creativity*, 165.

197   He wandered into the calligraphy: Steve Jobs, text and video of Stanford commencement address, June 12, 2005, posted on the Stanford Report, June 14, 2005, accessed September 4, 2012, http://news.stanford.edu/news/2005/june15/jobs-061505.html.

198   Without Peggy Guggenheim: http://en.wikipedia.org/wiki/Jackson_Pollock; http://totallyhistory.com/jackson-pollock/, accessed September 5, 2012.

198   in Long Island where his Drip paintings: http://totallyhistory.com/jackson-pollock/, accessed September 5, 2012.

198   Guggenheim also introduced: Helen Gent, "Peggy Guggenheim, Mistress of Modernism," *Marie Claire*, June 19, 2009, accessed September 5, 2012, http://au.lifestyle.yahoo.com/marie-claire/features/life-stories/article/-/5869429/peggy-guggenheim-mistress-of-modernism/.

198   Of course, Guggenheim played: Kay Larson, *Where the Heart Beats* (New York: Penguin Press, 2012), 96–97.

198   More recently, Stanford University: Liz Gannes, "Stanford Professors Launch Coursera with $16M from Kleiner Perkins and NEA," All Things D, April 18, 2012, accessed September 5, 2012, http://allthingsd.com/20120418/stanford-professors-launch-coursera-with-16m-from-kleiner-perkins-and-nea/; John Markoff, "Online Education Venture Lures Cash Infusion and Deals with 5 Top Universities," *New York Times*, April 18, 2012, accessed September 5, 2012, http://www.nytimes.com/2012/04/18/technology/coursera-plans-to-announce-university-partners-for-online-classes.html.

198   one of two VCs who invested: Markoff, "Online Education Venture."

198   In December 2009, Jesse Genet: http://www.kickstarter.com/projects/lumi/lumi-co-a-new-textile-printing-technology, accessed September 5, 2012.

199   a printing system based: Morgan Furst, Q&A with Jesse Genet, Source 4 Style, December 22, 2011, accessed September 5, 2012, http://www.source4style.com/trends/the-academy/qa-with-jesse-genet-printing-with-light/; http://lumi.co/, accessed September 5, 2012.

199   Angoulvant needed $50,000: http://lumi.co/collections/kickstarter, accessed September 5, 2012; http://www.kickstarter.com/projects/lumi/print-on-fabric-using-sunlight-the-lumi-process, accessed September 5, 2012.

199   Eddie Huang was a twenty-three-year-old: Baohaus story based on interviews with Evan Huang; Fresh Off the Boat, Eddie Huang's blog, http://thepopchef.blogspot.com/; *New York* magazine, accessed September 8, 2012, Salon, January 19, 2011; http://www.baohausnyc.com/about.html, accessed September 5, 2012.

200   the small meat-filled buns: Joe DiStefano, "A First Look at Baohaus (in Which I Learn I Fit the Profile), January 11, 2012, accessed September 5, 2012, http://newyork.seriouseats.com/2010/01/a-first-look-at-baohaus-review-lower-east-side-manhattan-new.html; http://www.baohausnyc.com/about.html, accessed September 5, 2012.

200   Huang wanted to call: Evan Huang, interviews with the author, spring 2012.

200   Eddie's father may have been: "The Year of Asian Hipster Cuisine," *New York* magazine "Grub Street," July 8, 2012, accessed September 5, 2012, http://newyork.grubstreet.com/2012/07/asian_hipster_cuisine.html.

200   Luckily, several of Eddie: Evan Huang, interviews with the author, spring 2012.

201   They arrived in 2009: interviews with Huang, spring 2012; http://thepopchef.blogspot.com/search?updated-min=2009-01-01T00:00:00-05:00&updated-max=2010-01-01T00:00:00-05:00&max-results=50, accessed September 5, 2012.

201   Baohaus catered the Kickstarter: Evan Huang, interviews with the author, spring 2012.

201   For about $4: Ligaya Mishan, "Baohaus," *New York Times*, February 23, 2010, accessed September 5, 2012, http://www.nytimes.com/2010/02/24/dining/reviews/24under.html.

202   Eddie and Evan use social media: http://thepopchef.blogspot.com/.

202   Evan even hired one: Evan Huang, interviews with the author, spring 2012.

202   The original Rivington Street: Evan Huang, interviews with the author, spring 2012; Josh Ozersky, "Meet Eddie Huang, Food Personality," *Time*, February 23, 2011, accessed September 5, 2012, http://www.time.com/time/nation/article/0,8599,2053195,00.html.

202   Eddie has worked with fashion: TSSCREW, "Hoodman Clothing," *Smoking Section*, December 1, 2008, accessed September 5, 2012, http://smokingsection.uproxx.com/TSS/2008/12/hoodman-clothing; Anne van de Sande, "Hoodman Clothing CEO Discussed Messages Behind New Line," *Baller Status*, July 15, 2008, accessed September 5, 2012, http://www.ballerstatus.com/2008/07/15/hoodman-clothing-ceo-discusses-messages-behind-new-line/.

202   that features illustrations criticizing: "In Defense of Chinese Dads," Eddie Huang, Salon, January 19, 2011, accessed September 5, 2012, http://www.salon.com/2011/01/19/in_defense_of_chinese_dads/.

202   On his blog "Fresh Off the Boat": http://thepopchef.blogspot.com/, accessed October 21, 2012; Eatocracy editors, "Chow 13 Honorees—A Sneak Peek," November 3, 2011, http://eatocracy.cnn.com/2011/11/03/chow-13-honorees-a-sneak-peek; TSSCREW, "Hoodman Clothing."

202   The Huangs have networked: Emily Nordee, "Talking with Eddie Huang," March 28, 2011, accessed September 5, 2012, http://www.foodrepublic.com/2011/03/28/talking-eddie-huang.

202   Eddie has been a guest: Matt Rodbard, "Eddie Huang Got a TV Show. Earned It," December 29, 2011, accessed September 5, 2012, http://www.foodrepublic.com/2011/12/29/eddie-huang-got-tv-show-earned-it; Allison Benz, "A Day in the Life of a Chef," Radio Blog, July 3, 2012, accessed September 5, 2012, http://theradioblog.marthastewart.com/2012/07/a-day-in-the-life-of-a-chef.html; Evan Huang, interviews with the author, spring 2012.

203   One incubator that nurtures: Y Combinator Site, http://ycombinator.com/, accessed September 5, 2012.

203   Architect Charles Gwathmey: http://www.trianglemodernisthouses.com/gwathmey.htm, accessed September 5, 2012.

204   There are also manufacturing platforms: Bob Brunner, discussions in author's class at Parsons; Tim Brown, discussions in author's class at Parsons.

204   NY Creative Interns: Emily Miethner spoke at my class and I interviewed her in the spring of 2012; http://www.hercampus.com/career/how-she-got-there-emily-miethner-founderpresident-ny-creative-interns.

204   Past events have included: http://nycreativeinterns.com/, accessed September 5, 2012; http://wearenytech.com/294-emily-miethner-founder-president-of-ny-creative-interns-community-manager-at-recordsetter-com, accessed September 5, 2012.

205   YouTube cofounder Chad Hurley: interview between Hurley and Bill Moggridge that I attended and participated in, 2011.

207   Brian Chesky, cofounder: Steven Loeb, "Airbnb Buys Up UK Rival, Crashpadder, Ahead of Olympics," VatorNews, March 20, 2012, accessed September 5, 2012, http://vator.tv/news/2012-03-20-airbnb-buys-up-uk-rival-crashpadder-ahead-of-olympics; Robin Wauters, "Airbnb Buys German Clone Accoleo, Opens First European Office in Hamburg," TechCrunch, June 1, 2011, accessed October 22, 2012, http://techcrunch.com/2011/06/01/airbnb-buys-german-clone-accoleo-opens-first-european-office-in-hamburg/.

207   And eBay has grown: http://news.cnet.com/2100-1017-941964.html.

207   Even Apple has begun: http://www.cultofmac.com/129150/this-norwegian-man-made-millions-selling-siri-to-steve-jobs/.

208   On January 28, 2010: http://www.economist.com/node/15393377, accessed September 8, 2012.

209   Prophet of Profits: I came up with this term in talking about the *Economist* cover with Ben Lee.

209   Charisma, secularized from: accessed September 5, 2012, http://oed.com/view/Entry/30721?redirectedFrom=charisma&.

210   "the quality of an individual": Max Weber, *The Theory of Social and Economic Organization*, tr. A. M. Henderson and Talcott Parsons (Oxford: Oxford University Press, 1947), 328, 358.

210   "The holder of charisma": Max Weber, "The Sociology of Charismatic Authority," from *Max Weber: Essays in Sociology*, tr. and ed. H. H. Gerth and C. Wright Mills (Oxford: Oxford University Press, 1946).

211   When Facebook "went public": http://stream.wsj.com/story/
        facebook-ipo/SS-2-9640/, accessed September 5, 2012.

211   But in all the conversation: http://www.sec.gov/Archives/
        edgar/data/1326801/000119312512034517/d287954ds1
        .htm#toc287954_10, accessed September 5, 2012.

211   In his IPO letter: Ibid.

211   To ensure that sentiment: Michael Hiltzik, "Facebook Sharehold-
        ers Are Wedded to the Whims of Mark Zuckerberg," *Los Angeles
        Times*, May 20, 2012, accessed September 5, 2012—http://arti
        cles.latimes.com/2012/may/20/business/la-fi-hiltzik-20120517.

211   Of course, the subsequent IPO: Roben Farzad, "Facebook:
        The Stock That Keeps on Dropping," *BusinessWeek*, August 2,
        2012, accessed October 15, 2012, http://www.businessweek
        .com/articles/2012-08-02/facebook-the-stock-that-keeps-on-
        dropping; Jessica Guynn, "Facebook: Mark Zuckerberg Won't
        Sell Stock for at Least One Year," *Los Angeles Times*, Septem-
        ber 4, 2012, accessed October 15, 2012, http://articles.lat-
        imes.com/2012/sep/04/business/la-fi-tn-facebook-ceo-mark-
        zuckerberg-wont-sell-stock-for-at-least-one-year-20120904.

212   From the 1920s through much: Rakesh Khurana, interviews
        with author; Roger Martin, conversations with author; Rakesh
        Khurana, *From Higher Aims to Hired Hands* (Princeton, NJ:
        Princeton University Press, 2007).

212   Back in 2004, when Sergey: http://investor.google.com/corpo
        rate/2004/ipo-founders-letter.html, accessed September 5, 2012.

212   "Sergey and I founded Google": Ibid.

213   Though Steve Jobs is now: Olivia Fox Cabane, "Can You Learn
        to Be as Charismatic as Steve Jobs?" Gigaom, April 25, 2012, ac-
        cessed September 5, 2012, http://gigaom.com/2012/04/25/can-
        you-learn-to-be-as-charismatic-as-steve-jobs/.

214   I saw Mark Zuckerberg: author's notes, World Economic Forum
        in Davos, 2009.

214   Mark Zuckerberg was once: Nicholas Carlson, "The Facebook
        Movie Is an Act of Cold-Blooded Revenge—New Unpublished
        IMs Tell the Real Story," *Business Insider*, September 21, 2010,
        accessed September 5, 2012, http://www.businessinsider.com/
        facebook-movie-zuckerberg-ims#.

214   Zuckerberg is a master at finding: Henry Blodget, "The Matura-

tion of the Billionaire Boy-Man," *New York* magazine, May 6, 2012, accessed September 5, 2012, http://nymag.com/news/features/mark-zuckerberg-2012-5/.

218   In her classic book: Julia Cameron, *The Artist's Way* (New York: Jeremy P. Tarcher/Putnam, 1992).

CHAPTER 8

223   In May and September 2012, Hewlett-Packard announced: Shara Tibken, "H-P Says It Plans 2,000 More Layoffs," *Wall Street Journal*, September, 10, 2012, accessed September 15, 2012, http://online.wsj.com/article/SB100008723963904441004045776433 52249358504.html; *New York Times* Business Day Companies, Hewlett-Packard Corporation (HPQ) News Report, August 23, 2012, accessed September 14, 2012, http://topics.nytimes.com/top/news/business/companies/hewlett_packard_corporation/index.html; Wendy Kaufman, "Hewlett-Packard Set to Lay Off 30,000 People," National Public Radio broadcast, May 18, 2012, accessed September 14, 2012, http://www.npr.org/2012/05/18/152979181/hewlett-packard-set-to-layoff-30000-people.

223   The decisions were made: Jordan Robinson, "Atop Meg Whitman's Worries: H-P's Size," *Associated Press*, September 23, 2011, accessed September 14, 2012, http://phys.org/news/2011-09-atop-meg-whitman-h-p-size.html;   Laurent Belsie, "Meg Whitman New HP CEO," *Christian Science Monitor*, September 23, 2011, accessed September 14, 2012, http://www.csmonitor.com/Business/2011/0922/Meg-Whitman-new-HP-CEO.-What-firm-has-more-CEO-change/Hewlett-Packard-4-CEOS.

223   Where once HP led the world: Stacey Vanek Smith, "Hewlett-Packard Reportedly Will Lay Off 30,000," *Marketplace Tech*, May 18, 2012, accessed September 14, 2012, http://www.marketplace.org/topics/tech/hewlett-packard-reportedly-will-lay-off-30000; Hewlett-Packard YouSigma SWOT Analysis: http://www.yousigma.com/comparativeanalysis/hewlettpackardswot.pdf, accessed September 14, 2012.

223   Even with a cutting-edge new technology: Ben Worthen, "H-P

Opts to Divest High-end Halo System," *Wall Street Journal*, June 1, 2011, accessed September 14, 2012, http://blogs.wsj.com/ digits/2011/06/01/h-p-opts-to-divest-high-end-halo-system; Larry Walsh, "HP Targets Cisco, Rivals in Networking Push," *Channelnomics*, December 9, 2011, accessed September 14, 2012, http://channelnomics.com/2011/12/09/hp-targets-cisco-rivals-networking-push/.

223   In the spring of 2012: One weekend in April 2012, I gathered together six former HP engineers in Portland, Oregon, to talk about the change in culture at HP over time. This was an extraordinary group of men who clearly loved working at HP for most of their careers and cared deeply about the company. Each one felt a deep sense of loss for HP's decline in innovation and creativity. I have omitted names to protect their identity, but I was moved by their insights and passion. Much of what follows emerged from that session (April 4 to 5, 2012).

224   Fiorina was brought in: Craig Johnson, "The Rise and Fall of Carly Fiorina: An Ethical Case Study," *Journal of Leadership and Organizational Studies*, November 2008, vol. 15, no. 2, 188–96.

224   Her strategy was to shift: Ibid.

224   While she was boosting: Larry Walsh, "Tale of Two Product Launches: Apple vs. HP," *Channelnomics*, January 16, 2012, accessed September 14, 2012, http://channelnomics .com/2012/01/16/tale-product-launches-apple-vs-hp/; John Martellaro, "Last Qtr: iPad Outsells HP's PCs," *Mac Observer*, January 25, 2012, accessed September 14, 2012, http://www.macobserver .com/tmo/article/last_qtr_apple_ipad_outsells_hps_pcs/.

224   once the tablet was introduced: Jack Schofield, "2010 PC Sales Grew 13.8%, but 2011 Looks Tough," ZDNet, January 13, 2011, accessed September 14, 2012, http://www.zdnet.com/2010-pc-sales-grew-13-8-but-2011-looks-tough-4010021485/; IDC Press Release, January 12, 2011, "PC Market Records Modest Gains During Fourth Quarter of 2010," http://www.idc.com/about/ viewpressrelease.jsp?containerId=prUS22653511&sectionId=null&elementId=null&pageType=SYNOPSIS, accessed September 14, 2012.

225   She was, as one former HP engineer: personal interviews with HP employees, April 4 to 5, 2012.

225   Decisions once made: Ibid.

225   New ideas were thereafter: Ira Kalb, "Everything at Hewlett-Packard Started to Go Wrong When Cost-Cutting Replaced Innovations," *Business Insider*, May 27, 2012, accessed September 14, 2012, http://www.businessinsider.com/heres-where-everything-at-hewlett-packard-started-to-go-wrong-2012-5.

225   Fiorina was pushed out: Joe Nocera, "Real Reason for Ousting H.P.'s Chief," *New York Times*, August 13, 2012, accessed September 14, 2012, http://www.nytimes.com/2010/08/14/business/14nocera.html.

225   He cut spending: personal interviews with HP engineers, April 4 to 5, 2012.

225   "Once you started worrying": Ibid.

226   But instead of looking: James B. Stewart, "Voting to Hire a Chief Without Meeting Him," *New York Times*, September 21, 2011, accessed September 14, 2012, http://www.nytimes.com/2011/09/22/business/voting-to-hire-a-chief-without-meeting-him.html.

226   After management consultants: personal interviews with HP engineers, April 4 to 5, 2012.

226   "We would see them only once": Ibid.

226   HP fights to remain: "PC Market Struggles, Lenovo Nearly Catches HP," CDR Info, October 11, 2012, accessed October 12, 2012, http://www.cdrinfo.com/Sections/News/Details.aspx?NewsId=34502; "HP Still Has Top Market Share for PCs," *Forbes*, October 13, 2011, accessed September 14, 2012, http://www.forbes.com/sites/marketnewsvideo/2011/10/13/hp-still-has-top-market-share-for-pcs/; Terrence O'Brien, "HP Reclaims Top Spot in PC Sales, Market as a Whole Climbs 21 Percent," engadget.com, May 1, 2012, accessed September 14, 2012, http://www.engadget.com/2012/05/01/hp-reclaims-top-spot-in-pc-sales-market-as-a-whole-climbs-21-pe/.

226   Even when its labs produced: http://www.polycom.com/products/polycom_halo.html, accessed September 14, 2012; Mark Speir, "Polycom Acquires HP's Halo Video Conferencing for $89M," RCR Wireless, June 1, 2011, accessed September 14, 2012, http://www.rcrwireless.com/austin/20110601/components/polycom-acquires-hps-halo-video-conferencing-for-89m/.

227   sociologist Erving Goffman: Erving Goffman, *Encounters: Two Studies in the Sociology of Interaction* (Indianapolis, IN: Bobbs-Merrill, 1961), 78; referenced in Clifford Geertz, *The Interpretation of Cultures* (New York: Basic Books, 1973), 436.

227   Great Recession would soon: Michael Lind, "The Failure of Shareholder Capitalism," Salon.com, March 29, 2011, accessed September 13, 2012, http://www.salon.com/2011/03/29/failure_of_shareholder_capitalism/; "A New Idolatry," *Economist*, April 22, 2010, accessed September 13, 2012, http://www.economist.com/node/15954434.

228   In May of 1970, Eugene Fama: Eugene F. Fama, "Efficient Capital Markets: A Review of Theory and Empirical Work," *Journal of Finance*, vol. 25, no. 2, Papers and Proceedings of the Twenty-Eighth Annual Meeting of the American Finance Association New York, N.Y., December 28 to 30, 1969 (May 1970), 383–417. Published by Wiley-Blackwell for the American Finance Association.

228   In it, Fama would take: Joe Nocera, "Poking Holes in a Theory on Markets," *New York Times*, June 6, 2006, accessed September 13, 2012, http://www.nytimes.com/2009/06/06/business/06nocera.html?pagewanted=all&_r=0.

228   In its purely financial form, EMT: Jeremy J. Siegel, "Efficient Market Theory and the Crisis," *Wall Street Journal* Online, October 27, 2009, accessed September 13, 2012, http://online.wsj.com/article/SB10001424052748703573604574491261905165886.html.

228   Of course, what was missing: I am indebted to Ben Lee, who received his PhD from the University of Chicago, for highlighting the difference between uncertainty and risk in the economic analysis and theory formation that came out Chicago's economics department. This distinction forms a major theme in the course we co-teach at Parsons. Ray Ball, "The Global Financial Crisis and the Efficient Market Hypothesis: What Have We Learned?" University of Chicago, *Journal of Applied Corporate Finance*, vol. 21, no. 4, 2009; Siegel, "Efficient Market Theory and the Crisis"; Roger Lowenstein, "Book Review: The Myth of the Rational Market by Justin Fox," *Washington Post*, June 7, 2009, accessed September 13, 2012, http://www.washingtonpost.com/

wp-dyn/content/article/2009/06/05/AR2009060502053.html.

228   "black swans": Nassim Nicholas Taleb, *The Black Swan: The Impact of the Highly Improbable* (New York: Random House, 2007).

228   By excluding uncertainty: Frank H. Knight, *Risk, Uncertainty and Profit* (New York: Sentry Press, 1921).

229   In the 1960s and 1970s, as EMT: John Cassidy, "The Minsky Moment," *New Yorker*, February 4, 2008, accessed September 13, 2012, http://www.newyorker.com/talk/comment/2008/02/04/080204taco_talk_cassidy.

229   Charles Kindleberger's: Charles P. Kindleberger, *Manias, Panics and Crashes* (Hoboken, NJ: Wiley, 1978).

229   British journalist and essayist: Walter Bagehot, *Lombard Street* (New York: Scribner, 1873).

229   And who hasn't heard: Charles MacKay, *Extraordinary Popular Delusions and the Madness of Crowds*, with a foreword by Andrew Tobias (1841; New York: Harmony Books, 1980).

230   The belief in the efficient market: Professors Roger Martin and Ben Lee pointed out the linkage between CEO pay, profits, and stock market performance to me; Michael Jensen and William Meckling, "Theory of the Firm: Managerial Behavior, Agency Costs and Ownership Structure," *Journal of Financial Economics*, October 1976, vol. 3, no. 4, 305–60, http://papers.ssrn.com/abstract=94043.

230   In their work, including a paper: Ibid.

230   Roberto Goizueta: "Coke CEO Roberto C. Goizueta Dies at 65," cnn.com, October 18, 1997, accessed September 13, 2012, http://www.cnn.com/US/9710/18/goizueta.obit.9am/;   Jerry Schwartz, "Roberto C. Goizueta, Coca-Cola Chairman Noted for Company Turnaround, Dies at 65," *New York Times*, October 19, 1997, accessed September 13, 2012, http://www.nytimes.com/1997/10/19/us/roberto-c-goizueta-coca-cola-chairman-noted-for-company-turnaround-dies-at-65.html.

231   After talking to Wall Street: interview with Professor Ho at a copresentation she gave with Gillian Tett of the *Financial Times*, March 10, 2010, at Columbia University; Karen Ho, *Liquidated: An Ethnography of Wall Street* (Durham, NC: Duke University Press, 2009).

231   The value of millions of houses remains: Binyamin Appelbaum,

"Cautious Moves on Foreclosures Haunting Obama," *New York Times*, August 19, 2012, accessed September 14, 2012, http://www.nytimes.com/2012/08/20/business/economy/slow-response-to-housing-crisis-now-weighs-on-obama.html; Les Christie, "Troubled Homeowners Get a Lifeline," *CNN Money*, October 24, 2011, accessed September 14, 2012, http://money.cnn.com/2011/10/24/real_estate/housing_refinance/index.htm.

231 Interest rates, zero for many: David Shulman, "The Downside of the Fed's Zero Rate Policy," *US News and World Report*, April 30, 2012, accessed September 14, 2012, http://www.usnews.com/opinion/blogs/economic-intelligence/2012/04/30/the-downside-of-the-feds-zero-interest-rate-policy.

231 the volatility of the markets: Tami Luhby, "Credit Freeze and Your Paycheck," *CNN Money*, September 28, 2008, accessed September 14, 2012, http://money.cnn.com/2008/09/28/news/economy/main_street_impact/index.htm?postversion=2008092811; Colin Barr, "How It Got This Bad," *CNN Money*, September 26, 2008, accessed September 14, 2012, http://money.cnn.com/2008/09/26/news/leverage.fortune; Martin Wolf and Chris Giles, "Transcript: Larry Summers Interview," *Financial Times*, April 2, 2010, accessed September 14, 2012, www.ft.com/intl/cms/s/0/3c023d9c-3dba-11df-bdbb-00144feabdc0.html#axzz24r1TJX1h.

232 It took years for economists: Marcus Baram, "Who's Whining Now? Gramm Slammed by Economists," September 19, 2008, accessed September 14, 2012, http://abcnews.go.com/print?id=5835269.

232 In a 2010 interview with Martin Wolf: Wolf and Giles, "Transcript: Larry Summers Interview."

232 Summers was, after all: Stephen Labaton, "Congress Passes Wide-Ranging Bill Easing Bank Laws," *New York Times*, November 5, 1999, accessed September 14, 2012, http://www.nytimes.com/1999/11/05/business/congress-passes-wide-ranging-bill-easing-bank-laws.html; Charles Ferguson, "Larry Summers and the Subversion of Economics," *Chronicle of Higher Economics*, October 3, 2010, accessed September 14, 2012, http://chronicle.com/article/Larry-Summersthe/124790/; Rana Foroohar, "Larry Summers: No Regrets on Deregulation," *Time Busi-*

ness, April 12, 2011, accessed September 14, 2012, http://
business.time.com/2011/04/12/larry-summers-no-regrets-on-
deregulation/.

232   In 1999, Summers, along with: Cyrus Sanati, "10 Years Later, Looking
at Repeal of Glass-Steagall," *DealBook*, November 12, 2009, accessed
September 14, 2012, http://dealbook.nytimes.com/2009/11/12/10-
years-later-looking-at-repeal-of-glass-steagall/.

232   the Depression-era regulation: Labaton, "Congress Passes Wide-
Ranging Bill Easing Bank Laws."

232   calling the repeal "historic": Sanati, "10 Years Later."

232   Perhaps no one believed: Justin Fox, "The Myth of the Rational
Market," *Time*, June 22, 2009, accessed September 14, 2012, http://
www.time.com/time/magazine/article/0,9171,1904153,00.html.

232   But in October 2008, Greenspan: Edmund L. Andrews, "Greenspan Concedes Error on Regulation," *New York Times*, October
23, 2008, accessed September 14, 2012, http://www.nytimes
.com/2008/10/24/business/economy/24panel.html.

232   "Those of us who have looked: Kara Scannell and Sudeep Reddy,
"Greenspan Admits Errors to Hostile House Panel," *Wall Street
Journal*, October 24, 2008, accessed September 14, 2012, http://
online.wsj.com/article/SB122476545437862295.html.

232   Greenspan was criticized: Andrews, "Greenspan Concedes Error
on Regulation."

233   Back in the mid-nineties: Katrina Brooker, "Citi's Creator, Alone
with His Regrets," *New York Times*, January 2, 2012, accessed
September 13, 2012, http://www.nytimes.com/2010/01/03/
business/economy/03weill.html.

233   On July 25, 2012: "Wall Street Legend Sandy Weill: Break Up
the Big Banks," CNBC Report, July 25, 2012, http://www.cnbc
.com/id/48315170, accessed September 13, 2012.

233   from 1977 to 2008: Roger Martin, "The Age of Customer
Capitalism," *Harvard Business Review*, January 2010, http://hbr
.org/2010/01/the-age-of-customer-capitalism/ar/1.

233   Compare that with: Ibid.

234   In 2009, Michael Mandel: I've known and respected Mike Mandel for all the years we worked together at *BusinessWeek* and the
years thereafter. We talk often about the issues of innovation and
growth that are discussed here, and I have written about Mike's

work in the past. Mike has the keen mind of an original thinker and a striking ability to see through the mass of data to find new patterns. He is a superb knowledge miner and dot connector. Mike is the most brilliant economist that I know. Michael Mandel, "The Failed Promise of Innovation in the U.S.," Businessweek .com, June 3, 2009, accessed September 14, 2012, http://www .businessweek.com/magazine/content/09_24/b4135000953288 .htm; Bruce Nussbaum, "America's Innovation Shortfall and How We Can Solve It," *Harvard Business Review*, September 20, 2011, accessed September 14, 2012, http://blogs.hbr.org/cs/2011/09/ americas_innovation_shortfall.html.

234   While futurists in the nineties: Michael Mandel, "My Review of Tyler Cowen's New Book," *Mandel on Innovation and Growth*, February 2, 2011, accessed September 14, 2012, http://innovationand growth.wordpress.com/tag/innovation-shortfall/; Michael Mandel, discussions with the author, 2012.

235   The consequences of this: Mandel, "The Failed Promise of Innovation in the U.S."; Michael Mandel, discussions with the author, 2012.

235   almost $7 trillion: Michael Mandel, discussions with the author, 2012; Ian Campbell, "U.S. August Net Trade Deficit Reported at U.S. $44.2 Billion," *Safe Haven*, October 17, 2012, accessed October 20, 2012, http://www.safehaven.com/article/27361/us-august-net-trade-deficit-reported-at-us442-billion.

235   Even in high tech: Michael Mandel, "Innovation Failure," *Mandel on Innovation and Growth*, October 5, 2010, accessed October 21, 2012, http://innovationandgrowth.wordpress.com/2010/10/05/ innovation-failure/; Michael Mandel, discussions with the author, 2012.

235   A National Science Foundation report: Mark Boroush, "NSF Releases New Statistics on Business Innovation," October 2010, accessed October 12, 2012, http://www.nsf.gov/statistics/infbrief/ nsf11300/.

235   At an Aspen Institute: TPI Aspen Forum, August 21 to 23, 2011, https://techpolicyinstitute.org/aspen2011/.

235   Peter Thiel, a cofounder: Rip Empson, "Max Levchin and Peter Thiel: Innovation in the World Today Is Between 'Dire Straits and Dead,'" *TechCrunch*, September 12, 2011, http://techcrunch

.com/2011/09/12/max-levchin-and-peter-thiel-innovation-in-the-world-today-is-between-dire-straits-and-dead/.

235   and we haven't seen any new breakthroughs in energy: Ken Bossong, "Renewable Energy Provided 11% of Domestic Energy Production in 2010," *Renewable Energy World*, accessed September 15, 2012, http://www.renewableenergyworld.com/rea/news/article/2011/04/renewable-energy-provided-11-of-domestic-energy-production-in-2010.

235   We have all been feeling the ripple: Michael Mandel, "Why Isn't the Innovation Economy Creating More Jobs?" *Mandel on Innovation and Growth*, February 22, 2012, accessed on October 21, 2012, http://innovationandgrowth.wordpress.com/2010/02/22/why-isnt-the-innovation-economy-creating-more-jobs-part-i/; Michael Mandel, discussions with the author, 2012.

236   As a consequence of the failure of this innovation: Francis Fukuyama, "A Conversation with Peter Thiel," *American Interest*, March/April, 2012, accessed September 14, 2012, http://www.the-american-interest.com/article.cfm?piece=1187; http://www.bls.gov/news.release/realer.nr0.htm, accessed September 14, 2012; Thiel's comments at the Aspen conference.

236   Young college grads: Michael Mandel, "The State of Young College Grads 2011," *Mandel on Innovation and Growth*, October 1, 2011, accessed October 21, 2012, http://innovationandgrowth.wordpress.com/2011/10/01/the-state-of-young-college-grads-2011/; Michael Mandel, discussions with the author, 2012.

236   To make matters worse: Michael Mandel, conversations with the author, 2012.

236   It isn't working for many: Michael Mandel, conversations with the author, 2012; US Census Bureau release, *Income, Poverty and Health Insurance Coverage in the United States: 2011*, accessed October 15, 2012, http://www.census.gov/newsroom/releases/archives/income_wealth/cb12-172.html.

237   "We need fewer efficient: J. Bradford DeLong, "Economics in Crisis," Project-Syndicate.org, April 29, 2011, accessed September 14, 2012, http://www.project-syndicate.org/commentary/economics-in-crisis.

237   In 2009, the futurist Paul Saffo: Paul Saffo presentation at the

HSM World Innovation Forum, May 4, 2009. Paul kindly granted me several interviews over the next few years. The following insights are informed by his presentation and materials from interviews with him.

238   In my fall 2012 Design: Personal record, 2012.

239   New companies (those less than five years old): Tim Kane, "The Importance of Startups in Job Creation and Job Destruction," http://www.kauffman.org/research-and-policy/the-importance-of-startups-in-job-creation-and-job-desctruction.aspx.

239   data that the Ewing Marion Kauffman Foundation: Ibid.

240   Building a new economics of creativity: This model came out of long discussions with Ben Lee and Mike Mandel and a great number of students in my classes at Parsons. I've put forth parts of the model in blogs on *Fast Company* and *Harvard Business Review* to generate discussion, but this is the first time I've tried to build the scaffolding of a cohesive paradigm of a creativity-driven capitalism.

242   Though he's best known: "University of Chicago Centennial Faculty Pages: Frank K. Knight," http://www.lib.uchicago.edu/proj ects/centcat/centcats/fac/facch23_01.html.

242   Chicago economist Frank Knight: Frank H. Knight, *Risk, Uncertainty, and Profit* (Washington, DC: Beard Books, 2002), 224–25.

242   In his book *Risk*: Knight, *Risk, Uncertainty and Profit*.

243   For Knight, social uncertainty: Ross Emmet, *Selected Essays*. Volume 1, "What Is Truth in Economics." Volume 11, "Laissez-faire: Pro and Con" (Chicago: University of Chicago Press, 1999).

244   Fortunately, there are "incubators": Nicole Davis, "Putting Your Money Where Your Mom and Pops Live," *Brooklyn Based*, March 3, 2012, accessed September 15, 2012, http://brooklynbased .net/email/2012/03/putting-your-money-where-your-mom-and-pops-live/.

244   As Amy Cortese, a crowdfunding: http://www.amycortese.com/ Amy_Cortese_homepage.html, accessed September 15, 2012; Davis, "Putting Your Money Where Your Mom and Pops Live."

244   "people need to be able: Amy Cortese, correspondence with author, August 7, 2012.

244   Cortese believes that: Ibid.

244   Smallknot, for example: Davis, "Putting Your Money Where Your Mom and Pops Live."

244   Egg restaurant in Brooklyn: Ibid.

244   The forty-five investors: Smallknot profile of Egg restaurant campaign, http://smallknot.com/egg, accessed September 15, 2012.

244   Smallknot was founded by two: Andrew Cominelli, "Smallknot Redefines Small Business Finance," *Greenpoint Gazette*, March 14, 2012, accessed September 15, 2012, www.greenpointnews .com/news/4316/smallknot-redefines-small-business-finance.

245   "We were working all hours": Ibid.

245   "Local crowdfunding can mitigate": Amy Cortese, correspondence with author, August 7, 2012.

245   In Britain, crowdfunding: http://locavesting.blogspot.com, accessed October 18, 2012.

245   Funding Circle, for example: http://www.fundingcircle.com/, accessed September 15, 2012.

245   If Americans shifted just half: Danielle Sacks, interview with Amy Cortese, "'Locavesting': Investing in Main Street Instead of Wall Street," *Fast Company*, accessed October 18, 2012, http://www.fastcoexist.com/1678356/locavesting-investing-in-main-street-instead-of-wall-street.

245   In *Capitalism, Socialism and Democracy*: Joseph Schumpeter, *Capitalism, Socialism and Democracy* (New York: Harper, 1975; orig. pub. 1942), 82–85.

CHAPTER 9

251   Sixteen of us had spent: Future of Design Summit, Stanford, March 19 to 20, 2010. Among the others at the summit were Nick Leon, now director of Design London; Bill Burnett, Executive Director of the Stanford Design Program; Ronald Jones, who leads the Experience Design Group at Konstfack University College of Arts, Crafts and Design in Stockholm; Nathan Shedroff, who was then launching a design strategy MBA at California College of the Arts in San Francisco; and Jacob Mathew, cofounder of IDIOM, a top innovation consultancy in India.

253   Roger Martin at the Rotman: I talk with Roger Martin often and have done many interviews and panels with him over the last

fifteen years, but he shared this important bit of advice about assessing creativity during an online discussion. Reena Jana, who was a member of our innovation and design team at *BusinessWeek* and is now Executive Editor at Frog, moderated this conversation between Roger and Dan Pink on March 18, 2011, on Educating the Creative Leaders of Tomorrow, put on as part of the Steelcase 360 Discussion series.

253   The Juilliard process: http://www.juilliard.edu/apply/program-information/dance/index.php#auditions, accessed September 6, 2012.

254   On *Dancing with the Stars*: http://beta.abc.go.com/shows/dancing-with-the-stars/about-the-show, accessed September 8, 2012.

254   A three-person panel of judges: Ibid.

254   In *The Creative Vision:* Mihaly Csikszentmihalyi and Jacob W. Getzels, *The Creative Vision: Longitudinal Study of Problem Finding in Art* (Hoboken, NJ: John Wiley & Sons, 1976).

254   Teresa M. Amabile later used: http://www.gifted.uconn.edu/nrcgt/reports/rm04202/rm04202.pdf, accessed September 6, 2012. As Keith Sawyer describes in *Explaining Creativity* (New York: Oxford University Press, 2012), 40–44, Kaufman, Amabile, and others in the field of creativity research favored what is called the consensual assessment technique (CAT) as a measurement of individual creativity.

255   In 2012, 7 percent of all: "100 Top MBA Employers 2012—Where MBA Students Said They'd Most Like to Work" http://money.cnn.com/news/economy/mba100/2012/full_list/, accessed September 8, 2012.

255   That placed the consultancy: Ibid.

255   On March 28, 2011, Brown: Tim Brown's presentation to my class was one of the most extraordinary talks on the history and future of design I have ever heard. It was simply brilliant. Tim explained the evolution of the field from designing products to social systems, from objects to interventions, from consumerism to social innovation. He pointed to start-ups and science as new frontiers for design, as well as health, education, and government itself. I've known Tim for over a decade and he's been both a pioneer in the field and an inspiration all this time.

256   Spotify, for example: http://www.glassdoor.com/Overview/

Working-at-Spotify-EI_IE408251.11,18.htm, accessed September 8, 2012; http://www.spotify.com/au/blog/archives/2012/08/29/can-you-solve-spotifys-tech-puzzles/, accessed September 8, 2012.

256   Continuum, the innovation: interview with Harry West, March 20, 2012.

258   a game called Odyssey: http://www.odysseyofthemind.com, accessed September 8, 2012. I have had a number of conversations about Odyssey of the Mind with Kelsey Meuse, a student of mine who played the game and loved it.

258   There are tournaments in cities: http://www.odysseyofthemind.com/learn_more.php, accessed September 8, 2012.

259   In the 2011 to 2012 game: http://www.odysseyofthemind.com/materials/2012problems.php?l=wf2012, accessed October 19, 2012.

259   A previous year's problem: http://www.odysseyofthemind.com/materials/2011problems.php#p4, accessed October 19, 2012.

259   The teams are then assessed: http://www.odysseyofthemind.com/whatis.php, accessed September 8, 2012.

259   Judges are mostly parents: http://www.capregboces.org/BOCESInsider/InsidePages/2010-11/01_21_OdysseyJudgesSIDELINK.cfm, accessed September 8, 2012.

259   Problem Captains: http://www.odysseyofthemind.com/judges/positions_descriptions.php, accessed September 8, 2012.

260   A small group of curators: www.kickstarter.com/discover/curated-pages, accessed September 8, 2012; Julia Cheng, "The Crowdfunding Festival—JOBS Act, Kickstarter & Indiegogo—What Do They Mean to Entrepreneurs?" July 29, 2012, accessed September 9, 2012, http://demystifylegal.blogspot.com/2012/07/crowdfunding-JOBS-kickstarter-indiegogo-crowdfunder.html.

EPILOGUE

263   In August 2012: E. S. Browning, Steven Russolillo, and Jessica E. Vascellaro, "Apple Now Biggest-Ever U.S. Company," *Wall Street Journal*, August 20, 2012, accessed October 22, 2012, http://online.wsj.com/article/SB10000872396390443855804577601773524745182.html; "Apple Becomes the Most Valuable Company Ever," CBS Money Watch, August 20, 2012,

accessed October 22, 2012, http://www.cbsnews.com/8301-505123_162-57496461/apple-becomes-most-valuable-company-ever/. In inflation-adjusted dollars, Microsoft was valued at about $850 billion in 1999, higher than Apple in August 2012. Many Wall Street analysts are predicting that Apple's stock would rise to $800 and perhaps even $1,000 over the next year or two, which would give it a market capitalization higher than Microsoft even when inflation is taken into account.

263   There are clear lessons: Japan's great moment of innovation in the 1970s and 80s came from a small group of post–World War II entrepreneurs, such as Sony's Akita Morita, who were not connected with the country's giant *zaibatsus*, or conglomerates. Morita was a hero to Steve Jobs. But Sony failed to integrate its digital and hardware operations to create a single engaging ecosystem and lost its lead to Apple. Japan's consumer electronics giants, once leaders, are now commodity manufacturers of TVs and other products. India, on the other hand, is proving to be very creative in developing new business models based on its traditional style or "frugal innovation."

264   That may be beginning to change: I've been to a design conference in Inchon put on by the mayor, whose two brothers graduated from RISD and Parsons. Parsons's largest alumni group is in Korea. I've also consulted once for Samsung by being on a panel that talked about design trends in front of the company's top designers.

264   In announcing its latest Five Year Plan: KPMG China, "China's 12th Five-Year Plan Overview," March 2011, accessed October 15, 2012, http://www.kpmg.com/cn/en/IssuesAndInsights/Articles Publications/Documents/China-12th-Five-Year-Plan-Overview-201104.pdf.

Index

About the Author

BRUCE NUSSBAUM, former assistant managing editor for *Busi-nessWeek*, is professor of innovation and design at Parsons School of Design and an award-winning writer. He is founder of the In-novation & Design online channel, and *IN: Inside Innovation*, a quarterly innovation magazine, and blogs at *Fast Company* and *Harvard Business Review*. Nussbaum is responsible for starting *BusinessWeek*'s coverage of the annual International Design Excel-lence Award and the World's Most Innovative Companies survey. He is a member of the Council on Foreign Relations. He taught third-grade science in the Philippines as a Peace Corps volunteer.